T0149065

SHADES
OF **BLUE**

Celebrating 35 Years of
Penguin Random House India

PRAISE FOR *CITIES AND CANOPIES*

'Packed with fun facts, engaging stories, and superb tales and factoids about Indian city trees'—Pradip Krishen, author, environmentalist and film-maker

'Just try this: walk by or be driven through any city, refusing flatly to look at any building or read any hoarding, with your gaze fixed only on the trees you pass. The impact is amazing. They embrace you, engulf you and transport you into the world of their fragile fascination and that of our life in nature. Harini Nagendra and Seema Mundoli do precisely that through this gripping journey into the world of trees, so close to us and yet so spurned by our cement minds and steel eyes. One closes this book with just one thought: thank you, trees, for just being what you are, where you are'—Gopalkrishna Gandhi, former IAS officer and former governor of West Bengal

'*Cities and Canopies* is a splendid book about trees and their many associates, ranging from birds and bats, wasps and ants to skinks and snakes, with fascinating titbits about science, such as how trees communicate with each other with the help of the fungal network connecting their roots; about history, such as how Kabir's great banyan tree on Kabirwad Island in Bharuch might in fact be much older and be the one that Alexander described; about how humans relate to trees, including their roles in games and art; their medicinal uses and how to concoct a jamun squash. A very enjoyable read that I highly recommend to all lovers of nature, culture and food'— Madhav Gadgil, ecologist, writer and columnist

'This book challenges the urban–rural divide in our minds. It helps city dwellers understand the dangers of nature-deficit disorder and rediscover the biophilia—a love for nature—that exists in us all. Harini Nagendra and Seema Mundoli do this with joy, professionalism and deep knowledge. They introduce us to the secret language of nature, such as the silent communication between the glorious amaltas tree and its carpenter bee pollinators. They frame interesting questions such as: can urban trees communicate with each other as well as those in forests? They also help us re-examine our animosity towards immigrant species, reminding us that the "sriphala" (the sacred coconut) and our very own tamarind are in fact exotics. A feast of a book, *Cities and Canopies* is timely and important for young people to read, and act upon'—Rohini Nilekani, philanthropist, journalist, author and social activist

'It's hard to read this book in one sitting because you have to take time off to deal with the flood of memories it triggers. Right from the first chapter, engagingly titled "A Khichri of Trees", the authors bring together scientific facts, history, trivia and personal experiences to outline how our cities can continue to coexist alongside trees; not in a preachy or didactic way but by outlining the connections that many Indian cities already have with what they describe as "nature's own museums"'—*The Hindu*

'It is a celebration of trees, a charmingly penned paeon to these great givers. I have little doubt that this occasionally idiosyncratic account, written with passion and a deep knowledge and affinity for trees, will make even those who scurry about their busy, regulated lives stop and look up at the canopies. And I think with awe, wonder, and humility'—*Hindustan Times*

'The book reveals myriad facts about the trees we've grown up watching and also gives us a painful reminder of their importance, given the tree-starved state of our urban landscape'—*Financial Express*

'Harini Nagendra and Seema Mundoli have given us a riveting work . . . I have not come across anything like it, connecting our past, present and future with trees'—Scroll.in

'The book is a luscious romp through the fruit, fun, poetry, folk tales, history and healing properties of the trees we live with. It also serves as a cautionary reminder that the magic we grew up with may not be evergreen after all. *Cities and Canopies* has the potential to transform you from a screen-time inspector to a parent who knows how to conjure up many hours of joy from seeds, fruits and flowers. It doubles up as an easy handbook to make your child appreciate and respect nature'—Livemint

'A delightful summer read, reminiscent of languid long summer afternoons, the book is everything you need to survive the hot summer—gentle, wise, peppered with tree-related trivia and more. In tree-starved cities that India now lives in, this book provides the much-needed cover—and of course, the impetus, for more trees. The book brings together urban stories of trees that dot the streets that need to be told and treasured. The writers, Harini Nagendra and Seema Mundoli, who are clearly tree lovers, tell them well'—*Week*

'In an age of instrumental views of nature centred around ecosystem services, these pieces of memory and frolic demonstrate that trees are ultimately beyond value'—Wire

'The authors bring attention to the natural elements of the urban environment that the next generation is going to grow up alongside and emphasise the need to protect it . . . With every chapter, the authors tap into the collective memory of a generation that grew up with trees. Nostalgia heavily supports the stories that take you back to memories you never knew you had . . . With an easy style of writing and by breaking down scientific facts into relatable bits of information, the authors make the book accessible to a wide audience'—Mongabay

SHADES OF BLUE

CONNECTING *the* DROPS *in* INDIA'S CITIES

Harini Nagendra
Seema Mundoli

PENGUIN
VIKING
An imprint of Penguin Random House

VIKING

USA | Canada | UK | Ireland | Australia
New Zealand | India | South Africa | China | Singapore

Viking is part of the Penguin Random House group of companies
whose addresses can be found at global.penguinrandomhouse.com

Published by Penguin Random House India Pvt. Ltd
4th Floor, Capital Tower 1, MG Road,
Gurugram 122 002, Haryana, India

Penguin
Random House
India

First published in Viking by Penguin Random House India 2023

Copyright © Harini Nagendra and Seema Mundoli 2023

Illustrations by Swati Chandak Sharma 2023

All rights reserved

10 9 8 7 6 5 4 3 2 1

The views and opinions expressed in this book are the authors' own and the
facts are as reported by them which have been verified to the extent possible,
and the publishers are not in any way liable for the same.

ISBN 9780670099696

Typeset in Sabon by Manipal Technologies Limited, Manipal
Printed at Thomson Press India Ltd, New Delhi

www.penguin.co.in

CONTENTS

ONE

BLUE WATERS

Water is the elixir of life, as we are often told. This magical liquid constitutes the medium in which life evolved. More than 71 per cent of the surface of the earth is covered by water, most of it in the oceans and seas that surround the continents. Water lies underground, in springs and reservoirs hidden from our sight. It fills the skies as water vapour, sits atop the oceans in large icy glaciers, and caps the surface of mountain peaks in the form of ice and snow. Water is even *in* us. Our body weight comes from water—at least 55–60 per cent of it. The clothes we wear, the coffee we drink, the food we eat, the table we use to write on—even the paper this book is made of, or the e-reader you are holding in your hand, have a water footprint. We would not exist today, if it were not for water.

How did the earth get its water? Earlier, scientists believed that when the earth formed, it was so fiery hot that any water that existed at the time would have been vapourized. They said that the water we see around us was carried to our planet from other heavenly bodies—planetesimals, comets and asteroids—which crashed into the earth after its formation. However, some recent research from the Carnegie Institution for Science, Washington, D.C., and the University of California, Los Angeles, suggests that the water we see around us may have existed during the early days of the earth's formation. Soon after the earth had formed, when it was still growing, its hydrogen-rich atmosphere may have interacted with hot magma, oxidizing to form water. Water on earth might even have multiple origins, as other recent research indicates.

Whatever its history, it is no accident that life evolved on earth, not on the other planets of the solar system. Living beings need water to survive, grow and reproduce. The earth is at the right distance from the sun. If the earth were farther from the sun, its water would be frozen, unavailable to support life. If it were any closer, the temperature would be scorching and the water would evaporate.

Even though most of the earth is covered with water, there is very little that we can access directly. Over 96 per cent of the earth's water lies in the seas and oceans. While the oceans support incredibly rich biodiversity, their waters are salty and cannot be directly consumed. We depend on freshwater, which constitutes only 3 per cent of the earth's water. Of this, 68 per cent is unavailable, locked up in ice, snow and glaciers. Only 0.5 per cent of the world's water is available to us as freshwater. We rely on this water for our survival.

Water has unique chemical and physical properties. One of these is a property we learn of in school—that it is a 'universal solvent'. Dihydrogen monoxide, the water molecule, contains two positively charged atoms of hydrogen and one negatively charged molecule of oxygen. When a water molecule encounters another compound, such as salt (sodium chloride), it dissociates into its constituent hydrogen and oxygen ions, which bind themselves to the ions in the salt molecule. Thus, salt dissolves in water. Water also dissolves oxygen, enabling fish and other marine life to breathe underwater using gills.

Pure water is so reactive that it can be harmful, pulling minerals out of our bodies, bones and teeth. If you leave pure distilled water in a glass bottle for a few weeks, it will begin to dissolve the container, leaving etchings on the inner surface. This is why bottled water manufacturers often leave a little bit of salt in the water they supply, satisfying the craving of the universal solvent for solutes and making it safer for us to drink. But this property of water is essential for life. Water dissolves most of the salts, minerals and other chemicals required to support life. As water moves through the bodies of animals and plants, it acts like a mini-food delivery and cleaning service, supplying food and nourishment to cells, and removing poisonous waste.

Water is also a 'sticky' liquid. Water molecules adhere strongly to each other, giving water the highest surface tension of all liquids (except for mercury). This property also comes from its polarity, because hydrogen and oxygen atoms of adjacent water molecules bond to each other. These bonds are especially strong at the surface, making the molecules cling together. The high surface tension of water makes soap bubbles round, delighting babies and adults alike. It is also responsible for the fact that the water droplets you see, running off a window in the rains, are spherical instead of elongated.

The high surface tension of water also has a biological function. It helps to support the biodiversity that lives on the surface of the water. Because of its stickiness, the water surface resists being broken or invaded by another object. Water thus acts like a supporting membrane, holding up many objects that are denser than it (as swimmers know). This is why you might see the leaves of many aquatic plants floating on the surface of lakes and ponds, supported by its surface tension. Water insects called striders also adroitly exploit this property to walk on the surface of lakes, trapping and feeding on flies and other insects.

Another important physical property of water is its high specific heat. It takes much more heat to raise the temperature of a given amount of water by 1°C compared to many other liquids. This is why the water stays cool even on a hot summer day, and rivers, lakes and ocean fronts help to cool down their surroundings. And why people throng to the beach in Mumbai on summer evenings—to take advantage of the cool sea breeze.

Water may be one of the most abundant molecules on earth, but little is fit for consumption. Most aquatic life thrives in a pH (potential of hydrogen) range of 6.5–9.0, while drinkable water lies in the pH range of 6.5–8.5. Yet factors such as acid rain and chemical pollution have altered the pH of many of our water bodies. Other factors that affect the quality and potability of water include its conductivity, the amount of dissolved oxygen, its turbidity, and the presence of disease-causing *Escherichia coli* (*E. coli*) and other microbial contaminants. Contaminated with industrial poisons, choked with sewage and garbage, the water in our cities is often toxic and unfit for drinking. Climate change will lead to further disruptions in water supply. Yet we fail to recognize the scale of the problem, or to act in ways commensurate with the magnitude of the disaster that looms on our horizon.

What would Chennai and Mumbai look like without their beaches? Would Delhi still be the capital city if the Yamuna River were to run dry? What if Bengaluru and Udaipur lost their lakes? Our cities would transform into deserts, unrecognizable and unliveable, without a drop of relief.

Civilizations have risen and fallen at the water's edge. The culture, history and daily life of Indian cities revolve around their water bodies. Rivers have inspired folk tales, myths, epics, paintings and operas. Across cities, people have mapped streams and swum in lakes, cruised on boats, picnicked on the beach, fished and farmed together, and fought pitched battles over water. There are many shades of blue in our ponds, wells, lakes and rivers. Each carries a different meaning, a different association. A *bhisti* (water carrier) in Delhi, a fisherwoman in a Bengaluru lake, a young boy who swims in the sea off the islands of Lakshadweep, a classical dancer in Udaipur who composes a dance opera to be performed by the lakeside and a mother in one of Mumbai's expensive flats who buys bottled water for her children—each is inextricably linked with water but in a very different way. The poor line up to purchase water in pots, rationing every drop— the wealthy purchase water by the gallon, using it to fill swimming pools, musical fountains and artificial waterfalls.

We cannot live without water. But it seems we cannot live *with* water either. Claiming to worship our rivers, we fill them with garbage and poison them with industrial effluents. Attending water conservation workshops in luxury hotels, we sit in large, air-conditioned rooms, becoming dehydrated and refreshing ourselves with gulps of bottled water. In doing so, we are heedless of the water footprint of such events and the bottling plants that supply them with water, or the fact that the people who live around these areas go thirsty.

Our relationship with water seems almost schizophrenic. One of the most common descriptions of our relationship with water is that it is like a life-giving parent to us. Like Maa Yamuna, as one of the boatmen on the river in Delhi told us. A mother we take for granted, who is ever-forgiving, ever-nurturing, sacrificing everything for us. But that analogy can only be stretched so far. Few of us would treat our mother the way we treat our lakes and rivers, poisoning, choking and smothering them with our waste.

As two residents of Bengaluru, we—like the readers of this book—have an intimate relationship with water. In writing this book, we draw on our own extensive body of research, which examines the relationships between people and water across Indian cities and rural areas, as well as the research of many others who have worked on these topics for decades. Although reasonably comprehensive, this book does not aim to be an exhaustive guide to issues of water management in cities. That would result in a massive, multi-volume series that you would need to wheel around on a pushcart. Instead, we offer an eclectic sampler, an idiosyncratic journey that takes you through the many avatars and aspects that water assumes in our daily lives. You will be introduced to fantastic monsters and water beasts, learn the names of spies who mapped rivers, read about the controversial history of big dam projects and river transportation, and dip into tales of songs, saints, ghosts and real-life water warriors. Alongside, we shine a spotlight on cities like Delhi, Bengaluru, Kolkata, Udaipur and Mumbai, highlighting their long history of association with water, a relationship which shaped their past and continues to influence their future.

We hope that the narratives we share in the pages of this book, collated from stories of research and action across our cities and from around the world, bring a deeper understanding of the many facets of blue water in our daily lives. Through stories of joy and sorrow, greed and selfless care, dismay and hope—for there is always hope.

Our desire, above all, is to offer you a *fun* book, a book which you can dip into and out of, to satisfy your thirst for knowledge about water and to whet your appetite for more. We believe that knowledge is power. Our hope is that the stories in this book will inspire you to learn more and to ask what we can do, individually and collectively, to make things better.

TWO

DELHI: RE-IMAGINING THE YAMUNA

The sacred Yamuna originates in the Yamunotri glacier, high up in the Himalayan mountains, where the river water is cold, pure and crystal clear. The waters gush down the mountains, tumbling over rocks and boulders, until the river meets its first impediment at the Tajewala Barrage, 172 km from its origin. After this, the river never regains its original purity. By the time the Yamuna reaches the capital city of Delhi, it has become a toxic cocktail of waste.

Amoeba-like, the boundaries of Delhi have shifted shape. Over the centuries, various rulers established at least nine ancient and medieval cities in Mehrauli, Siri Fort, Firozabad, Shahjahanabad, Shergarh, Quila Rai Pithora and nearby sites. Despite a dizzy dance of changing allegiances, the Yamuna River remained a constant fixture in the lives of the city's residents, supplying the city with water through a network of tributaries and streams, canals, stepwells and tanks.

One of the oldest such water-holding structures in Delhi is Surajkund. This tank was built by Tomar Rajputs in the eighth century, along with Badkhal Lake. Hauz-i-Shamsi, the largest water tank in Delhi, was built in the thirteenth century by Sultan Iltutmish (father of Razia Sultana, the only woman monarch of Delhi). In the fourteenth century, Firoz Shah Tughlaq, the king of the Delhi Sultanate, diverted the waters of the Yamuna through a canal directly into his fort, Firoz Shah Kotla, thus ensuring that the wells in the fort never ran dry. These channels were later maintained and repaired by the Mughal kings Akbar and Jahangir, and further extended during Shah Jahan's reign. By 1843, Shahjahanabad had 600 wells which supplied the city with water. Small streams were connected to ponds and tanks, and used to recharge groundwater, which then fed the wells and their larger counterparts, the *baoli*s of Delhi.

By 1803, the city belonged to the British. The expanding colonial city needed an ever-increasing supply of water. The water bodies of Delhi were failing, becoming polluted and falling into disrepair. The Ali Mardan Khan Canal, constructed during Shah Jahan's reign to conduct water from the Yamuna to the Red Fort, was an engineering marvel. Fed by the Yamuna and the Sabi rivers, the canal supplied many nearby tanks with water. By the mid-eighteenth century, this once-glorious canal lacked water. Although the British administration restored the canal in the early nineteenth century, they did not understand the importance of the connection between the canal and its tanks, leaving the tanks dry. In 1846, another tank was built to supply drinking water to Delhi, but its water became brackish and unfit for drinking within just a few years. Writing to his friend Yusuf Mirza in 1859, the famous poet Mirza Ghalib spoke bitterly about the destruction of his beloved city by the British.

> From Ammu Jan's Gate to the moat of the Fort, except for Lal Diggi and one or two wells, no trace of any building will remain . . . I should rejoice in the desolation of Delhi. When its residents have gone, then to hell with the city.

On their part, the British worried about the impact of traditional practices of night soil (human and animal faeces) disposal, in sewers and pits. Toxic discharge from these pits contaminated the subsoil, making its way to the Yamuna, giving rise to epidemics like cholera. Advancements in science and technology spurred their interest in finding technological solutions to the problem of water scarcity. By the mid-nineteenth century, advancements in water chemistry led to the development of quantitative tests to assess water quality. These tests convinced British administrators that much of Delhi's drinking water was impure, unfit for consumption.

Bombay (now Mumbai) and Calcutta (now Kolkata) had moved to a piped water system to ensure the supply of clean water. Buoyed by this success, the British government directed the Delhi Municipal Committee (DMC) to develop a similar plan. In 1869, the DMC developed a proposal to supply piped water to Delhi, conveying it from unpolluted streams, canals and wells outside the city's periphery. Although initial attempts were not very successful, by the beginning of the twentieth century, Delhi had a piped water network in place. Unfortunately, the city had also grown substantially by then. The waterworks system, sufficient for the needs of the nineteenth-century city, could only provide half the water the city required by the time it was built. The project also came with its social costs. An entire hamlet, the village of Chandrawal, had to be removed and relocated to make way for the project.

As the city grew, so did its appetite for water. By 1911, the British moved their capital from Calcutta to Delhi. They searched for a suitable location to build the new capital, one that could showcase the majesty of the British Empire. After rejecting the east bank of the Yamuna because it was too swampy, and the west bank because it was already developed, they settled on an area that lay between the Yamuna and the central ridge of the Aravalli Hills. This became Lutyens' Delhi. To supply the new capital with water, they built an additional pumping station on the Yamuna, but the shortage of water persisted.

Alongside water supply, the quality of water continued to be a cause for concern. Alum was used to precipitate impurities and chlorine to purify the water. Yet contaminated sewage continued to make its way into Delhi's water supply, causing frequent outbreaks of cholera and enteric diseases. Administrators began to restrict a number of activities which they termed 'unsanitary'. In 1917, the British administration in Delhi banned the cultivation of melons within 1.6 km of the New Water Waterworks in Wazirabad, compensating the farmers for their loss of income. Failing to take the sociocultural realities of local relationships with water into account, they prohibited activities like bathing and washing in the river, banning the use of the riverbank by *dhobi*s (washermen/washerwomen). At the same time, they began to view the riverfront as a place of recreation, making plans for parks and a golf course. Although these plans did not materialize, a new imagination of the Yamuna came into play under the colonial administration. The prohibitions they put in place began a process of exclusion of local communities, distancing them from the river—a process whose signature is still visible.

After a major flood in October 1955, the Yamuna River altered course, moving closer to the Najafgarh Channel. Once a part of the Sabi River, which flows down from the Aravalli Hills to Delhi, the Najafgarh

Channel feeds into Najafgarh Lake. When the Yamuna shifted close to the sewage-contaminated waters of the Najafgarh Channel, the river waters became contaminated, affecting the water supply of Delhi and causing a jaundice outbreak.

Despite pumping crores of rupees into cleaning the Yamuna, the situation has worsened. The Yamuna Action Plan-II was sanctioned in 2005, and the Delhi Jal Board initiated a plan to set up 60 km of drains in 2007. By 2018, the city had spent more than Rs 1500 crore to clean the river, to little effect. Though the Yamuna passes through Delhi for just 22 km (less than 2 per cent of the entire river's course), it picks up about 80 per cent of its pollution load from the city. After the Wazirabad Barrage, where water from the Yamuna is collected and purified for drinking, the Yamuna collects polluted water from twenty-two drains. The waste carried by the river has almost doubled between 1982 and 2019. No wonder the polluted river continues to froth, spreading toxic foam across the region.

More than half of the Yamuna basin in Delhi is cultivated. Melons, strawberries, cucumbers, ladies' fingers and other produce make their way from the banks of the river into Delhi's vegetable markets. Farmers also own livestock that graze at the edge of the river, drinking from its polluted waters. The toxic chemicals in the vegetables, meat and milk from this landscape are consumed by the city's residents, impacting their health. Many of these farms have been locally owned for decades. Migrants from many parts of India come to work in the fields. The farms contribute to the local economy, providing livelihoods for many.

River pollution now threatens these livelihoods. As one farmer we interviewed said, 'Earlier, we could grow crops in huge quantities. Pollution in the river has affected the fertility of the land. Our agricultural fields will become barren in the next fifteen to twenty years.'

Other, more niche livelihoods such as swimming and diving have become equally threatened. To prevent drowning accidents on crowded festival days, the British government used to employ divers. In 1945, as recorded in the Department of Delhi Archives, the Hindustan Scouts Association wrote to the Delhi Improvement Trust asking for 2222 sq. yards of land on the banks of the Yamuna to set up a swimming and rescue club to train divers. The chairman of the Trust had responded, quite presciently, 'Ultimately, a great many of these charitable trust requests for land are intended at worst as fraudulent means of obtaining land for private possession at a cheap rate.' (While this may have been true, it was especially ironic coming from an official of the British Empire, which itself grabbed a great many lands, in different parts of the world, for private possession at a cheap rate.)

The association did not get the land it asked for, but many expert swimmers trained in the river. Even today, such swimmers are kept on-call to prevent drowning accidents, especially during festivals that attract huge crowds. They work on a contract basis, supplementing their meagre income from the river through private lessons in swimming pools. Yet it is their work in the river that they are most proud of. By saving people from drowning, they believe they are fulfilling a sacred responsibility given to them by Maa Yamuna.

The boatmen we interviewed on the lake expressed similar sentiments. Just a few decades back, they took people on boat rides along the banks of the Yamuna, or farther out in the river to perform the last rites of the dead. With the increase in pollution, boating has drastically reduced, impacting their income. Some of the boatmen have found unique ways to derive an income from the river. An elderly boatman plies his boat to collect trash, such as cans, bottles and bags from the water, selling them to make money. He believes that this gives him an opportunity to stay connected to the river. Another uses the river to stage fashion photography and pre-wedding photoshoots, creating an Instagram account to popularize his work. He said, 'Through this idea [of organizing photoshoots] Maa Yamuna has provided me with new opportunities after the death of my father.'

In a city to which migrants flock in the lakhs, hoping to make ends meet, the riverbank does not disappoint. The banks of the Yamuna offer refuge to the homeless, who shelter on its banks and under its bridges. Devotees bring food to the poor at the riverside, as thanksgiving for a new car or a new job. Children who live near the ghats dive into the Yamuna, collecting coins thrown into the river by devotees. They also collect plastic bottles, clothes and temple waste, selling them for meagre profits. Coconut shells, blankets, pillows, even the clothes on the dead—anything left on the banks of the river is recycled, becoming a source of income for another. Even the hair that devotees discard, when they tonsure their heads for rituals, is not thrown away. Waste collectors scavenge and sell the hair to wig makers. In doing so, they perform a service to the city, by cleaning the river, yet their contributions remain unrecognized.

Along the ghats, adjacent to the temples, are educational institutions, *akhada*s (wrestling pavilions), small shops and businesses. Some, like the Nigambodh Ghat, are used for cremation, with steps leading into the Yamuna to perform the last rites of the deceased and immerse the ashes in the river. Some priests and residents who live along the ghats are saddened by the change they see. They say cremation has now become a 'business', performed without regard for the holy task. But for many of the priests, the river continues to provide a link to their past. Their forefathers told them about the special taste the Yamuna gave to dal cooked in its waters, a taste they can no longer experience.

Along the banks of the Yamuna, lamps lit on the ghats that line the river shine like jewels against the waterfront, creating a magical appearance—if one does not look at the garbage floating in the river, or take in the stench. Despite the visible evidence of pollution, devotees persist in their faith and respect for the river. On Chhath Puja, a day dedicated to the Sun God, worshippers immerse themselves in the polluted river, heedless of the toxic foam that makes its way into their eyes and lungs. As a ghat owner who takes a dip in the river each day, before performing his daily *aarti* (prayer), said, 'I am aware of the pollutants in the water, but it will always remain Goddess Yamuna for me.' A priest added, 'Just by the sight of Yamuna and Ganga, all sins are washed off—drinking and bathing in the river are secondary.'

Not everyone believes without question. Some residents say that fewer devotees come to the ghats these days, because of the condition of the river. Many devotees refrain from ancient traditions of *aachman* (taking in sacred river water through one's hands as a holy offering), wary of the sewage in the water, and the possibility of falling sick.

Meanwhile, heedless of the condition of the river, commercial riverfront development, first visualized by the British, is visible everywhere. These include urban mega projects such as the Akshardham Temple and the Commonwealth Games Village Sports Complex (today a gated community). Informal settlements and slums were demolished to make way for urban redevelopment, rendering tens of thousands of people homeless. Poignantly, many who lost their homes attributed this to Maa Yamuna's anger.

What does the national capital of India, Delhi, give to the sacred Yamuna? Sewage, industrial waste, toxic foam and urban projects that constrict its flow. And what does Delhi get in return? Water for the city, a place for swimming, sacred ghats, a space to grow food and a riverfront for recreation—all this and much more.

Like the ever-generous tree in Shel Silverstein's children's classic *The Giving Tree*, Maa Yamuna gives generously, without consideration of the one-sided nature of her relationship with the capital city. Once central to the life and growth of Delhi, the Yamuna now seems to have disappeared from its very consciousness and imagination.

Clash of Cultures: Scientific and Ritual Ideas of Water Purity

By the mid-nineteenth century, British 'scientific' ideas of water purity began to collide head-on with Indian caste-based notions of purity. These clashes were exacerbated by the advent of piped water technologies. The colonial regime used chemical tests to declare that piped water was clean and pure. Missing from these definitions were issues of caste and power, traditionally accorded to terms like 'pure' and 'impure' in India. In the nineteenth century, ideas of water purity were firmly intertwined with practices of untouchability, and rarely questioned. Night soil was once manually extracted from toilets of caste households by manual scavengers from oppressed castes who were considered 'untouchable' by privileged castes. Manual scavenging is an abhorrent, evil practice that continues despite being declared illegal, still leading to deaths across the country.

By 1866, Bombay and Calcutta had begun to use piped water, as did other colonial cities like Lahore. Planners selected an uncontaminated source of water outside the city, transporting drinking water to the city through a network of pipes and passages made of stone, brick and iron. The British considered this to be the safest way to supply pure drinking water, but some Indian residents were not convinced. In Calcutta, Brahmin groups complained that the waterworks was managed by labourers from 'lower' castes. They fretted about losing their caste status if they drank the water supplied by the British, saying this would make them ritually impure. Other, more scientifically minded Indians mocked their fears. Eventually, scriptural workarounds were found. Over time, people became accustomed to these new technologies. Public resistance to the use of mechanically purified filtered water reduced, and piped water became widely accepted in these pioneer cities.

THREE

WALKING ON WATER

In the south-western city of Stavanger, Norway, a new experiment is underway, to convert the local diesel-powered ferry to a high-speed electric version. Billed as 'carbon-free transport', the boat, named *MS Medstraum* (which means 'with electricity' in Norwegian) will be powered by a rechargeable electric battery. The boat is revolutionary in design. It is made using lightweight aluminium to reduce the energy required to move it across the water. The design is modular, with a view to reusing much of the structure of a more conventional ferry, such as the passenger seats, to maximize cost-effectiveness and sustainability. If all goes well, the company developing the ferry, Kolumbus AS, aims to expand this approach across Europe.

Another start-up, SeaBubbles, is working on developing a hydrogen-powered hydrofoil, a flying water taxi. These water taxis, or boats, hover above the surface of water when going at high speeds to reduce friction, reducing energy consumption. Manufactured in France, these boats will be available for commercial use in Europe soon.

The European Union has committed to reducing carbon emissions from transport by 90 per cent by 2050. Low-carbon methods of transport on water can help Europe meet its climate mitigation targets. They could be helpful in India as well, where 13.5 per cent of the carbon dioxide emissions come from the transport sector. Currently, 90 per cent of India's transport sector emissions are from road transport. But many decades ago, before the British introduced railways to India, rivers and canals were the main mode of inland transportation. Powered by human effort, this was a low-carbon form of transport long before the term was invented.

Historical documents describe boats, plying on rivers across the country, carrying people and goods from one place to the other. Chanakya's famous third-century treatise on governance, *Arthashastra*, laid out rules for travellers on boats. All travellers by ferry were to be inspected, looking out for suspicious passengers such as:

> One who has stolen the wife, daughter or wealth of another;
> one who acts suspiciously or appears agitated;
> one who carries too heavy a load;

one who disguises himself as an ascetic;
an ascetic who has erased the signs of asceticism;
one who pretends to be ill [to avoid paying the fare];
one who is panic stricken;
one who carries secretly valuable objects, letters,
    weapons, fire-making equipment or poison;
one who has travelled a long way without a pass.

Abul Fazl's *Ain-i-Akbari*, a detailed account of administration under the Mughal emperor Akbar, says that a boatman who navigates the treacherous rivers must have specific capabilities. He needs to be 'an imposing and fearless man, must have a loud voice, must be capable of bearing fatigue, active, zealous, kind, fond of travelling, a good swimmer'. Boatmen were paid wages, and there were specific rates for transport of different goods and animals, but the poor travelled free.

In southern India, coracles, round and lightweight, were used more widely. The Portuguese writer and traveller Domingo Paes, visiting the city of Vijayanagara in the sixteenth century, described coracles as basket-boats made of cane and leather. Each coracle could carry fifteen to twenty passengers. Even today, visitors to the ruins of Hampi, the capital of the Vijayanagara Empire, can ride in these coracles, floating on the Cauvery River.

Transportation on water was faster and cheaper. People preferred this approach to travelling on bullock carts on the roads. But apart from the larger rivers like the Ganga and the Brahmaputra, most other rivers were narrow and often silted. They could only accommodate smaller boats, which moved slowly, and carried light loads of passengers and cargo.

The colonial era saw the further development of inland navigation through steam-powered ships and boats. The British, interested in moving goods and military troops across the country, began to invest their energies in mapping the network of rivers in India. They first focused on the rivers in the north and the east. James Rennell, the British Surveyor General of Bengal, in his *An Account of the Ganges and Burrampooter Rivers* says that the two rivers and their tributaries 'form the most complete and easy inland navigation that can be conceived'. The Narmada, with its rocks, shallow stretches and rapids, was more difficult to navigate but formed another important waterway, helping move commodities from central India to the coast.

Calcutta, on the banks of the Hooghly River, was an important port city for the British Empire. The river was marked by an especially treacherous stretch near the Bay of Bengal, making it dangerous for ships to reach the port. The force of the river tide or a treacherous current could send boats off course, grounding them, even leading to accidents when boats collided. Much money and effort was spent to deepen the river, increasing the water level by diverting water from other rivers into the Hooghly. When none of these worked to satisfaction, the British constructed a series of canals, including Tolly's Nullah, built by William Tolly in 1777, which connected the Hooghly in the west to the Bidyadhari River in the east. Another canal, the Circular Canal, enabled the movement of goods and people between Calcutta, eastern Bengal and Assam.

The advent of steam technology was a game changer. By 1787, the American inventor John Fitch had built and demonstrated the utility of a 45-ft-long steamboat on the Delaware River, in USA, patenting his invention in 1791. In 1807, Robert Fulton demonstrated the first commercially viable steamboat, launching the 150-ft-long boat on the Hudson River to travel from New York City to Albany. In 1812, Fulton approached the East India Company, offering to develop steamboats for commercial trade and military use. Many early steamboats were dangerous as the boilers could explode. Technology quickly advanced, and their use became widespread. Previously, ships moved using wind power, through sails. The boats had to be made of wood as the wind could not move anything heavier, and this limited the size and speed of boats. Steamships could be made of iron, a sturdier material. They could be much larger, and move much faster, because of the increased energy from steam that propelled their movement.

It took just a few years for this new technology to reach India. The Nawab of Awadh commissioned a small steamboat for his personal use. The first river steamboat built in India was the *Diana*, made with Indian teakwood in Calcutta, and launched in 1824.

By 1825, the steamboat *SS Enterprise* had travelled all the way from the shores of England to the coast of India, with the Nawab of Awadh financing 20 per cent of the cost. The trip was supposed to take seventy days but ended up taking 114! Nevertheless, it was a promising beginning. The *SS Enterprise* was immediately purchased by the East India Company for the vast sum of Rs 40,000 and used during the Anglo-Burmese Wars. Like its later fictional namesake, the *Starship Enterprise* of the television series *Star Trek*, the *SS Enterprise* had made history, boldly going where no man had gone before.

Following the success of steamship use for ocean travel, the British turned their attention to inland river navigation by steam vessels in Calcutta. The first two vessels, *Brahmapooter* and *Hooghly*, were built by the Howrah Docks Company. The *Hooghly*, 102-ft-long and 18-ft-broad vessel built of teak, went from Calcutta to Allahabad (now Prayagraj). The *Brahmapooter* travelled from Calcutta to Assam on its first journey in 1828. In 1834, a regular steamer service connected Calcutta to Allahabad.

Tea was brought by steamships that ran from Guwahati to Calcutta in 1847. In 1856, another route was introduced from Guwahati to Dibrugarh, located farther east. From here, Assam tea was exported to the international trade market. By facilitating the movement of commodities such as cotton and tea, steam navigation helped fuel the growth of the city of Calcutta, which was the second most important urban centre of the British Empire by the end of the nineteenth century.

Indian businessmen, including Dwarkanath Tagore (Rabindranath Tagore's grandfather) invested in new steamers, sensing a new opportunity to grow their investment. An ocean-going steamship was built at the Bombay Docks, the ship-building yard run by the Wadias, Parsi industrialists. The *Hugh Lindsay* embarked on its first journey from Bombay in 1829, taking twenty-one days and eight hours to arrive in Suez. Later, each steamship built at the Mazagon Docks in Bombay carried a silver nail hammered into the wood, a traditional Parsi blessing for good luck.

Despite the interest expressed by Indian royalty such as the Nawab of Awadh, industrialists like the Wadias and businessmen like Dwarkanath Tagore, who wanted to invest in these new technologies, the British government maintained a tight control over these investments. The East India Company and private British investors sought to maintain a monopoly over river trade (as they did with the railways).

Initially, steamships were expensive, and only used to transport select commodities such as tea, as well as for military purposes. In time they became less expensive, transporting passengers as well as a larger variety of commercial goods. By the 1850s, the Oriental Inland Steam Company and the Indus Steam Flotilla plied their steamers in the rivers of the Punjab (in present-day Pakistan). The Ganga and the Indus were crowded with boats at one point. But the success of the inland steamship was short-lived, collapsing with the advent of the Indian railways. By the late nineteenth century, most transport was conducted by rail. Very few steamers were seen on the Ganga and Indus, though they continued to ply along the Brahmaputra and its tributaries, including in eastern Bengal (Bangladesh). Even today, in Bangladesh, steamships continue to play a key role in the transport of cargo.

A few decades after Independence, the idea of using rivers for transportation received a boost with the establishment of the Inland Waterways Authority of India in 1986. In 2016, the National Waterways Act identified 111 stretches of rivers and canals that could be used for shipping. The Inland Vessels Act 2021 was also passed 'to promote economical and safe transportation and trade through inland waters'.

Using rivers as waterways requires much less fuel than road, air or train travel, but rivers tend to silt, and it is difficult and expensive to maintain them as waterways. Traditionally, river transport was propelled by human power, by boatmen and women rowing the boat. This was a less invasive approach that enabled boats to coexist with aquatic biodiversity. But diesel-powered boats pollute river water and air, disturbing the ecology of the river. Some of the newly expanded large waterways pass through national parks and sanctuaries that are important habitats for severely endangered species, possibly disturbing their movement.

While the national project to use waterways for transport has not taken off at scale, heritage canal systems are being revived by local citizens in some cities. Kerala has an intricate system of waterways that connects the coast to the interior. The Connolly Canal in Calicut was built by M.V. Connolly, the then collector of Malabar. Opened in 1848, it linked the Korapuzha River in the north of the state to the Chaliyar River in the south and connected Calicut to Buddagherry (now Vadakara). The canal was used to transport rice from the south to the north of Kerala, reducing the risk of travel by sea. It linked the northern parts of the state to trade networks. The canal promoted the trade of spices in Calicut, contributing to the prosperity of the region. Today, the canal is filled with urban waste and sewage. However, after the floods of 2018, efforts have been initiated to clean the Connolly Canal. Perhaps the canal can reclaim its past glory as a heritage spice route for visitors. We can only hope.

The Kakinada-Puducherry stretch, connected by the Buckingham Canal, is another famous canal that makes its presence felt in the city of Chennai—by way of its stench. Restoring this would also help reduce flooding in the city, simultaneously helping to restore urban heritage.

Boats are largely associated with pleasure rides today, though a few cities continue to use water for transportation. For instance, Kolkata continues to have an active ferry service across the Hooghly, and ferries still run in parts of Kerala and Goa, connecting smaller secluded islands to larger towns. Some of these ferries are large enough for a car to be loaded on to them, and inexpensive too. The Hooghly ferry service enables commuters to enjoy a scenic, quick trip across the river for the grand sum of Rs 6. Tourists weary of pounding the streets of Kolkata can also hire a hand-rowed boat and float along the Hooghly, taking in the sight of the Howrah Bridge and the picturesque ghats. The Yamuna River, though now polluted and stinking, is also used for boating by families and groups of young people. Of course the experience of boating on the Yamuna today is nothing like the trips on pleasure boats enjoyed by nobles during the Mughal times.

In Bengaluru, as in many other south Indian cities, local fishermen continue to use coracles even today to navigate to the centre of the lake and cast their nets. In the floods of August 2022 when many apartments and gated communities in Bengaluru were submerged, coracle boats came to the rescue, helping evacuate residents and their pets to safety.

If only humans could walk on water! But perhaps this is not for mere mortals.

In the New Testament the Gospel of Matthew, Mark and John famously describe Jesus walking on water. In the Iddhipada Vibhanga Sutta, a Buddhist canon describing the teachings of Gautama Buddha, it is said that a monk who attained the four *jhana*s (states of spiritual power) could gain various supranormal abilities including the capacity to walk on water. A story of Padmapada, the famed devotee of the Vedic scholar Adi Shankaracharya, described how the Shankaracharya once called to him on the other side of the Ganga. As he started walking towards his guru, he absent-mindedly stepped on to the river. Lotuses (*padma*) appeared under his feet (*pada*) ensuring he did not drown, giving him his name.

With the application of a little ingenuity and the use of new technology, perhaps humans *can* walk on water. Dwarka Prasad Chaurasia, an inventor from Mirzapur, had built a cycle that can float on water, using an air-filled float with empty cans of ghee to give it buoyancy, and moving the float by adding a propellor to the back wheel. In his seventies, this senior entrepreneur later found a way for people to stay afloat and move on water using specially made footwear from thermocol and rexine and oars of thermocol. Using this, people could walk upright on the water, maintaining their balance on the shifting waves.

Walking on water on floats made of thermocol seems cumbersome, but using floating cycles, hydrogen-powered hovercraft and non-polluting electric boats could be ideas worth exploring.

With a little thought, and investment in Indian innovation, could we, like Norway and France, dream of reclaiming our waterways as carbon-free, non-polluting, biodiversity-friendly modes of transport?

Monopolizing the Ocean

The British monopoly over inland transport and railways extended to ocean transport. By the early twentieth century, one company—the British India Steam Navigation Company (BISNC)—controlled almost all the trade in the Indian Ocean. Started in 1856 by two Scotsmen, William Mackinnon and Robert Mackenzie, as the Calcutta and Burmah Steam Navigation Company to connect Calcutta with Rangoon (now Yangon), this company later merged with BISNC, also started by Mackinnon.

In the south Indian city of Tuticorin (now Thoothukudi), the influential lawyer and freedom fighter V.O. Chidambaram Pillai created one of the first indigenous shipping companies in India, the Swadeshi Steam Navigation Company, to oppose the British monopoly on ocean trade. Before this, the traders of Tuticorin were forced to use ships owned by BISNC to trade with Colombo. Pillai was a member of the Swadeshi movement and raised money from Indian businessmen in various parts of the Madras state to lease ships from a firm in Bombay, Essaji Dodgibhoy. The British government applied pressure on the Bombay firm to cancel the lease. Undeterred, Pillai then leased a ship from Sri Lanka and purchased two of his own ships. They arrived in Tuticorin from France in 1907, and he outfitted the ships with flags that proudly proclaimed 'Vande Mataram'.

Despite the best efforts of Pillai, the Swadeshi Steam Navigation Company was not successful. The British gave free tickets to their ships, reducing the profitability of Pillai's ships. Pillai was sent to jail in 1908 for his involvement in the Tuticorin Coral Mills strike, forced to do harsh imprisonment that included being harnessed to an oil press like a bull. His shipping company closed in 1911, forcing him to sell one of its ships to BISNC. Pillai was released from jail in 1912, but his law licence was revoked, leaving him destitute until it was later restored. His later years were difficult, and he died in 1936. To honour the revolutionary spirit of this brave industrialist, the Thoothukudi port was later renamed V.O. Chidambaranar Port Trust.

The Champion of Steamships

The British engineer Sir Arthur Cotton was a vigorous champion of the steamship. He believed that the waters of India were more valuable than the gold of Australia, because of their use for irrigation and navigation. In 1854, he laid out the advantages of transporting cotton from Berar Province to Bombay by water. By bullock cart, this journey would take close to seventy days to reach the Bombay port, from where the cotton was taken by ship to the mills of Manchester. This was not only slow but expensive, with frequent losses due to damage in the rains. He calculated that if the cotton was transported on the Godavari River, they could earn £2,10,000 every year on 20,000 tonnes of cotton.

Cotton estimated that more than 8000 km of rivers could be used for communication and navigation, drawing up a list that included the Ganga, Yamuna, Gogra, Chambal, Gandak, Soane, Bhagirathi, Mahanadi, Narmada, Krishna, Godavari, Wardha and Manjira. To him, railroads were a 'complete delusion' when it came to connecting a large country like India. River navigation could do the same at a fifth or even a tenth of the cost. By 1911, it was clear that Cotton's visions for river development were not going to move ahead. He left India a sad man, seeing his plans in ruins.

If Cotton's vociferous defence of waterways had been successful, how would transport in India seem today? Would we be booking steamer tickets instead of railway tickets?

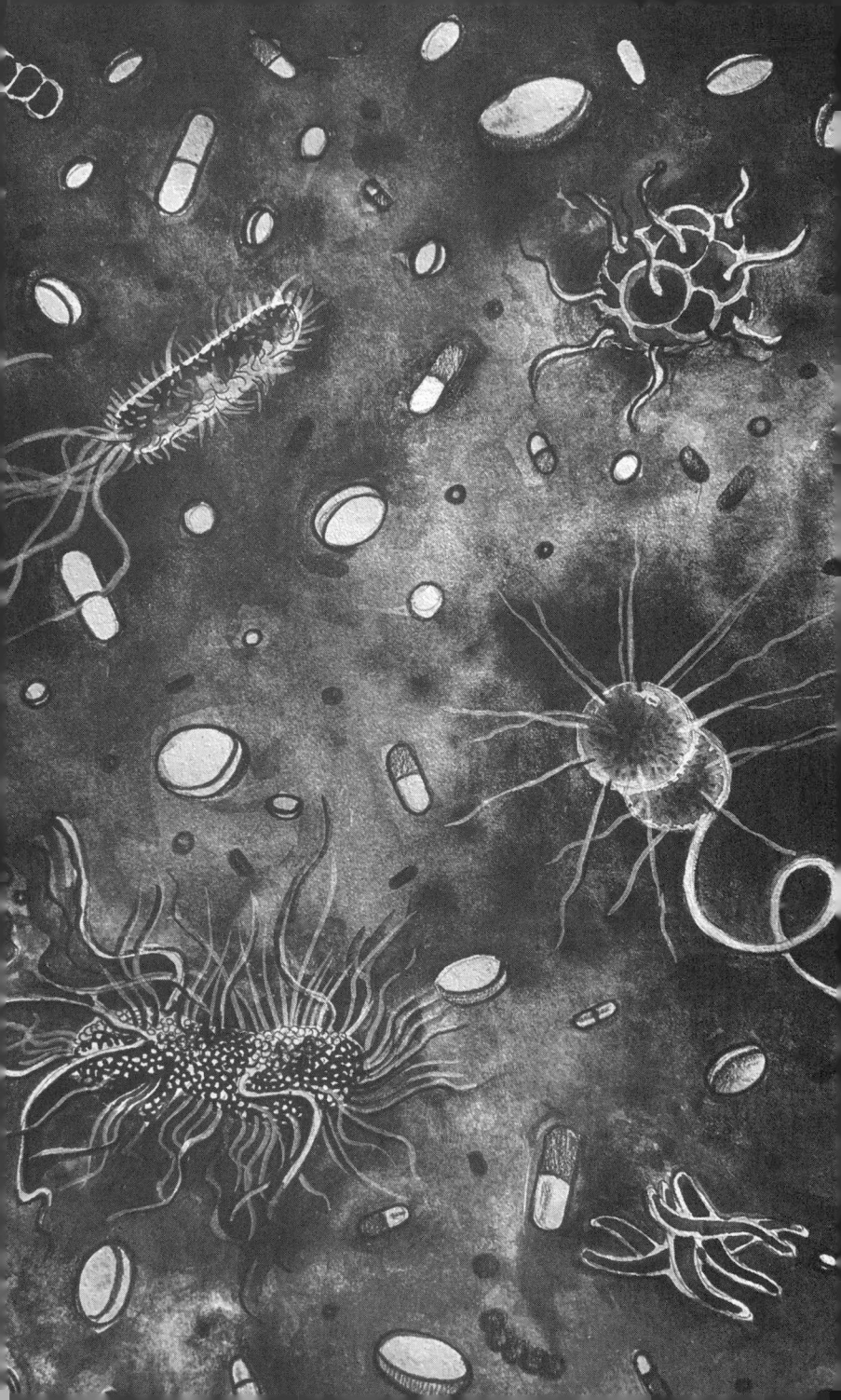

FOUR

OF DRUGS AND SUPERBUGS

The thoughtless person playing with penicillin is morally responsible for the death of the man who finally succumbs to infection with the penicillin-resistant organism. I hope this evil can be averted.

Writing in the *New York Times*, Alexander Fleming, who won a Nobel Prize for his discovery of antibiotics, cautioned the world about the dangers of antibiotic resistance as far back as 1945. With resistance to antibiotics spreading in cities because of drug contamination in lakes and rivers, the world is now waking up to the true significance of Fleming's prescient warning.

An accidental discovery on his return from a summer holiday in 1928, Fleming's discovery of penicillin, the miracle bacteria-killer, changed the world of medicine forever. Despite the excitement around the drug, he was aware that it was not a miracle cure capable of infinite use. Antibiotics could quickly become useless if microorganisms were exposed to small doses, too low to kill them all but enough to stimulate natural selection, so that antibiotic-resistant bacteria would survive. Fleming gave speeches around the world, cautioning about the possibility of resistance to antibiotics. Despite his messages of caution, the first case of penicillin resistance was recorded in 1945. Since then, cases of antibiotic resistance have been recorded across the world.

Antibiotic resistance, or more broadly antimicrobial resistance, occurs when microbes (which include bacteria, viruses, fungi and parasites) are exposed to small amounts of antimicrobial drugs such as antibiotics, antifungals and antivirals. Some microorganisms may develop resistance to many different kinds of antimicrobials. They become 'superbugs', bugs with superpowers of resistance. Like Kryptonite, the only mineral that can stop the invincible Superman, we need to find the one drug that can combat these superbugs. These microbes are isolated from patients, cultured in the lab and tested against various antimicrobials, to identify the drugs that can kill them. These medicines are then used to control the spread of the superbug in affected patients.

The first superbug gene from India was isolated from an Indian resident in Sweden, a fifty-nine-year-old man who fell ill after visiting India in the winter of 2007. Hospitalized in Ludhiana and then in Delhi, he was given a combination of five different antibiotics—amoxicillin and clavulanic acid, as well as metronidazole, amikacin and gatifloxacin. While still recovering, he left India and was admitted in a nursing home in Sweden. The doctors treating him in Sweden recognized that the bacteria causing his infections carried a gene that was responsible for antibiotic resistance. They isolated this gene from a plasmid in the disease-causing bacteria.

What are plasmids? Apart from their chromosome, many bacteria carry additional genes on small circles of DNA called plasmids. Plasmids can make copies of themselves independently and jump with ease from one kind of bacterium to another. When a gene for antibiotic resistance develops on a plasmid, it can therefore spread quickly to other bacteria. Managing plasmid-transmitted antibiotic resistance is a challenge for intensive care units in hospitals, where superbugs can lead to patient death if not addressed in time.

A follow-up journal article published in the well-known medical journal *Lancet* named the gene coding for antibiotic resistance 'New Delhi metallo-beta-lactamase 1'. The article said that the gene was also found in bacteria from several patients in India and Pakistan, and in people in the UK who had recently returned from these countries. The authors warned people against seeking medical treatment in India or Pakistan where antibiotic resistance was widespread. Britain issued an advisory warning its citizens against going to India for medical tourism. The Indian Council of Medical Research and the Ministry of Health raised objections to the article, calling it irrational. The editor of *Lancet* later apologized for naming the superbug after New Delhi, admitting it was an 'error'.

Despite the controversy around its discovery, this superbug gene spread fast, and is now found in more than a hundred countries. It was recently found in the Arctic, one of the most remote parts of the world, indicating that antibiotic resistance is now a truly universal challenge that leaves no part of the globe untouched.

Not all microorganisms are harmful—the vast majority of microbes which we encounter are harmless, or even helpful. Microorganisms such as bacteria, fungi, viruses and parasitic worms (helminths), too tiny to be seen with the naked eye and visible only under microscopes, can be found in all kinds of plants and animals. An estimated 10–100 trillion microbial cells are found in the human body. Living in symbiosis with us, most microbes assist us in essential functions such as digestion, enable the production of vitamins such as B12 and potassium in our bodies, help degrade toxins and strengthen our immune system. Some live as commensals, sharing space in our body without causing us harm. These commensals can even contribute to a couple of kilogrammes of our body weight. But some microbes are harmful—parasites depleting our resources for their benefit, or pathogens causing diseases that can debilitate and even kill us. Antimicrobial drugs, which function by either directly killing microorganisms or by reducing their capacity to multiply and grow, have been life savers.

Who amongst us has not popped an antibiotic pill or capsule? Ciprofloxacin for an upset stomach, Amoxicillin for a sore throat, Erythromycin at the first sign of a fever—anything to put us out of our immediate misery. The common practice of taking antibiotics for simple routine illnesses has increased our collective, societal likelihood of developing antibiotic resistance. Many types of factory-farmed meat, such as battery hens and aquacultured fish and shrimp, are routinely treated with antibiotics. Drug-enriched animal waste from these factories and farms is dumped in landfills, entering ponds, lakes and rivers and increasing their antibiotic load. Discarded medicines also find their way into ill-managed city landfills, and the antibiotic-rich leachate seeps into the soil, further contaminating urban water bodies. Sewage from our homes and hospitals contains antibiotic residues, which wastewater treatment systems cannot remove. These also enter the water systems around the city.

India is one of the world's largest manufacturers of antibiotics. Many pharmaceutical companies that produce cheap antibiotics in bulk have been accused of dumping their effluents directly into adjacent rivers and water bodies without adequate treatment. In 2007, Joakim Larsson, a Swedish researcher, analysed the effluents from a wastewater treatment plant in the city of Hyderabad that received water from ninety bulk drug manufacturers. The treated effluent contained concentrations of Ciprofloxacin that were astoundingly high, over a thousand times more than the recommended human dose. This antibiotic-rich effluent drains into a series of local streams, eventually flowing into the Godavari River. Originating in Maharashtra, the Godavari goes through breathtakingly beautiful valleys and lush green paddy fields before reaching the Bay of Bengal. How much toxicity does it carry along with it, and what kinds of drug resistance is it spreading as it passes through these rice fields and into the ocean? The implications are frightening.

The Ganga and Yamuna rivers also contain a toxic cocktail of wastes that harbour antibiotic resistant microbes. So does the Mula–Mutha River of Pune, which has the antibiotic-resistant *E. coli.* Although found to be naturally distributed in the environment and in our gut, some pathogenic strains cause food poisoning as well as other infections that can be dangerous. A common disease-causing bacterium in India, it is alarming to find *E. coli* becoming resistant to antibiotics. Many other rivers in southern India including the Cauvery, Tamraparni and Vellar are contaminated with antimicrobial drugs. The Pichavaram mangroves of Chennai, the Dal Lake in

Srinagar, the *kund*s (holy ponds) of Ayodhya and the springs of Sikkim also contain drug-resistant microbes.

Water is supposed to be the elixir of life. But in our rivers today, it has turned into a toxic cocktail of drugs and superbugs, resembling *kalakuta*, the poison drunk by Lord Shiva during *samudra manthan* (the churning of the Ocean of Milk).

When the water is polluted, the fish cannot be far behind. Five species of commonly eaten fish in Mumbai's harbours—Indian mackerel, false trevally, Indian oil sardine, goldspotted grenadier anchovy and tigertooth croaker—had high levels of antimicrobial-resistant pathogens because of the untreated waste from homes, industries, pharmaceutical companies and agriculture that is dumped into the sea. A study of the common bottlenose dolphins in Florida's Indian River Lagoon identified antibiotic-resistant bacteria from nose swabs. What about the endangered Ganges River dolphin of India, found in the polluted Ganga? We can only speculate.

India is not unique in facing a superbug problem. In a global study conducted in 2019, scientists from the University of York looked for the presence of fourteen commonly used antibiotics in rivers across the globe. Worryingly, 463 of the 711 locations they studied across seventy-two countries showed the presence of antibiotics. Iconic rivers such as the Tigris, Tiber, Thames, Danube and Mekong have been contaminated by antibiotics. But there is an especial urgency in the case of India, which is sometimes called the 'antibiotic capital of the world', both for the volume of drugs we produce as well as the quantity we consume. Newspaper headlines such as '"Superbugs" kill India's babies and pose an overseas threat' or 'India's sepsis babies: The struggle to save newborn babies from superbugs' are sensational but betray a fear that is all too real. This is especially worrying for the poor in cities, many of whom live close to polluted water bodies with greater exposure to superbugs while they lack access to good-quality, affordable healthcare.

Antimicrobial resistance has been aptly described as a 'tragedy of the commons'. While the consumption of antimicrobials benefits each of us as individuals, the burden of overuse and misuse of antimicrobials is borne by society. This is our collective problem, for which we need a collective solution. Patients, doctors, hospitals, elected representatives, policymakers and the healthcare industry must collaborate towards this.

In Hindu mythology, the Vaitarani River is believed to separate the earth from *naraka* (hell). The souls of the dead are believed to have to cross this river, filled with blood and crocodiles. Going by the state of our rivers today, even this mythological river may be filled with drug-resistant microbes. For good health *or* good karma, shouldn't we stop dumping our untreated wastes into rivers, lakes and ponds?

FIVE

MUMBAI: WRESTING LAND FROM THE OCEAN

In the popular imagination today, Mumbai is the Maximum City—a city always running at high speed. But as J. Gerson da Cunha, a Goan physician and historian, says in his book *The Origin of Bombay* (1900), the city was 'nothing more originally than a group of small islands, with numerous breakwaters, producing rank vegetation, dry at one time, and at another time overflowed by the sea'.

Mumbai is built on land reclaimed from the sea. A cosmopolitan city, Mumbai is a global financial hub that is home to some of the richest and the poorest Indians. Underneath the glitz and grime, unnoticed by many, lie the watery foundations of the islands on which the city is built. Over centuries, the history of the city's growth, a process driven by capital, industry and the construction boom, has been erased from collective memory.

The area that is now Mumbai city was originally a group of seven islands: Bombay, Mahim, Colaba, Old Woman's Island (Little Colaba), Worli, Mazagaon and Parel, which were connected during low tide but separated by the waters of the sea during high tide. Ptolemy, the legendary Greek astronomer, mathematician and geographer, named an archipelago of seven islands Heptanesia in 150 CE. Many believe that he was speaking about today's Mumbai. The first inhabitants to settle in this region are believed to be the Kolis, a fishing community that still inhabit the Koliwadas or fishing villages of the city.

The Portuguese acquired the islands from the Mughals in 1530, at a time when only about 10,000 people lived in this region. The British East India Company received the islands as part of the dowry of Catherine Braganza, daughter of the king of Portugal, who married King Charles II of England in 1661. They found the islands to be an ideal place to build a protected harbour for trade.

The British tried to map the collection of islands that marked the boundaries of the archipelago. This seemed impossible. The boundary of the coastline shifted with changes in the tide, and with the season. The islands were surrounded by swampy, saline marshes with a network of creeks that looked solid but could not support the growth of the city. To fulfil the aspirations of the British for aggressive expansion, the sea had to be pushed back. The marshy wetlands between the islands, along with their network of creeks, were termed 'wastelands'. They were considered to be places of sickness, where diseases like cholera and malaria claimed many thousands of lives. Bombay was even called Yamapuri, the abode of death. Such language paved the way for reclamation. It seemed not only acceptable but essential, the right thing to do to reclaim the marshes, civilizing and sanitizing the growing city.

The long process of wresting the city from the sea was initiated by the British. At first, they thought of filling in the partially submerged land between the islands. The first proposal for land reclamation was made in 1668, but it moved at a snail's pace. Much effort was expended in trying to fill in the Great Breach, the creek between Mahalaxmi and Worli, in the early eighteenth century, but this proved difficult, taking several decades to be completed successfully. In 1727, the low-lying swamps of Mahalaxmi were filled in and reclaimed for agriculture. The fort in Dongri and the hill on which it was located were demolished in 1769. The material from the demolition was used to fill up the low-lying land in the vicinity. The process of land filling and reclamation thus proceeded in fits and starts across the larger landscape until the Hornby Vellard project, under which a series of causeways were constructed between 1782 and 1845. This finally connected the seven islands into one landmass, the city of Bombay.

Reclaiming the city required quantities of material. The hills around the city were reduced to rubble and used to fill in the creeks, but this was an expensive process. The frugal Bombayites found another approach, reusing the waste of the city. The Elphinstone Dock was built in 1858 by filling a creek with the garbage of Bombay, a clever scheme that brought down the cost of construction. 'Bombay owes everything to successive reclamations', as a 1926 report of a government-appointed committee to look into the reclamation of Back Bay aptly put it.

Now called Mumbai, the urban district is an island today hemmed in by the Arabian Sea, Thane Creek and Harbour Bay. The suburban district of Mumbai, which extends farther towards the inland, has scope to grow, but away from the favoured areas that lie along the coast. Meanwhile, because of its history of reclamation, most of the land surface of Mumbai is just a few metres above sea level, making it prone to repeated flooding, particularly during the monsoon.

The landmark, devastating flood of 2005 is unforgettable for most Mumbaikars. But events like Cyclone Nisarga, which hit the coast near Mumbai in 2020, are set to become a regular feature of the Mumbai landscape under climate change, making much of the city's landmass unliveable for some time in the year. Oblivious of the watery future that awaits it, the city defiantly continues to plan for infinite growth, on a finite patch of vulnerable land.

At greatest risk are the mangroves, a fascinating coastal ecosystem containing trees uniquely formed to survive on the interface between land and sea. Mangrove trees have several unique adaptations that enable them to survive underwater. Their filtration system allows them to take in the salty seawater and remove the excess salt. Their aerial stilt roots help the trees to stand upright and resist the pressures of the changing tides, with pneumatophores, lateral roots that grow out of the mud, helping the trees take in oxygen. Mangroves are critical coastal ecosystems and an important habitat for fish to breed and reproduce. They are extraordinarily biodiverse, home to animals as varied as insects, birds like the flamingo, reptiles such as the saltwater crocodile, and mammals that include bats and monkeys. Mangroves are a critical buffer against climate change, protecting the coast from extreme weather events such as storm surges, cyclones and tsunamis.

Mumbai's mangroves cover an area of 65 sq. km including Manori and Malad creeks on the west, and Thane Creek on the east, which has the longest mangrove patch. The fishing Koli community has a close association with the mangroves of Mumbai. Living in forty-five fishing villages spread across Mumbai, they once supplied the Bombay Duck, or bombil, a favourite fish of Mumbaikars that is found in abundance locally.

Describing the fishing of the Bombay Duck in 1863, Govind Narayan, a well-known author of that time, writes in his autobiography *Govind Narayan's Mumbai*: 'Thousands of khandies of dried bombil are consumed in Mumbai annually. The Kolis take their boats to fish in the sea around Mumbai up to a distance of around sixty kos. They use large nets.'

The coastal creeks of Mumbai also provided an ideal place to harvest salt, from saline ocean water, evaporating in the shallow creeks under the harsh glare of the sun's rays. With urban growth, the salt pan workers and the Kolis, iconic to the city's history and heritage, are under threat, as is the coastline of Mumbai. The Coastal Regulation Zone Notification of 1991 and the subsequent notification of 2011 offer legal protection to the mangroves, salt pans and coastal ecology, and to dependent communities such as the Kolis. The Sewri Mangrove Park was set up in 1996 to protect 15 acres of mudflats and mangroves between Sewri and Trombay. A Mumbai High Court Order in 2005 decreed a complete stop to destruction and cutting of mangroves and banned construction within a 50-m radius of mangroves. Koliwadas and Gaothans, villages where fishing communities reside, are classified as 'no development zones' in the Mumbai Metropolitan Region.

Yet, urban growth continues to impact local communities and coastal biodiversity. Seventy acres of the Mithi River's estuary was reclaimed to set up the Bandra–Worli Sea Link, while a Special Economic Zone of 137 acres is proposed to be built on the Thane-

Mulund Creek in an area occupied by mangrove forests. The Chhatrapati Shivaji Maharaj Smarak is near an important fish-breeding site. The Navi Mumbai International Airport is located in an area with mangroves and mudflats. The 21.8-km Mumbai Trans Harbour Link will impact the Thane Creek Flamingo Sanctuary, disturbing the thousands of migratory greater and lesser flamingos that visit the creek every year between December and March and turn the entire place a glorious pink. Oil pipelines and oil rigs have polluted the sea, affecting coastal biodiversity.

A number of initiatives across Mumbai, spearheaded by different groups, offer hope to its residents. The architect and urban designer P.K. Das has helped to restore Mumbai's beaches and coastal parks as urban commons, accessible to students, Kolis, wealthy residents, tourists and vendors. A young lawyer and environmental activist named Afroz Shah started what came to be recognized as the world's largest beach clean-up movement in Mumbai in 2015. Beginning with a small section of the beach, his efforts have grown into a beach clean-up movement that has attracted contributions from Kolis, schoolchildren, slum residents, politicians and Bollywood movie stars. In 2018, after the removal of tons of trash, the restored sandy beaches of Versova witnessed a magical sight, the return of hatchlings of the endangered Olive Ridley marine turtle, which had not been seen at the site for nearly two decades.

Marine Life of Mumbai is a citizen's collective that documents the flora and fauna along the coastline. The group conducts shore walks that reveal the magical world of marine life nestled amidst the rocks and tide pools. Children and adults interact with marine life such as hermit crabs, sea anemones and clams that they have only seen in animation films. These walks provide unforgettable experiences that foster environmental awareness, creating more advocates for the protection of Mumbai's coasts. The Ministry of Mumbai's Magic, another collective, also works towards raising awareness about Mumbai's biodiversity. In April 2021, they launched the 'Making Art for Mumbai's Mangroves' programme inviting contributions in different forms—digital, watercolour painting and embroidery—about mangroves and the communities inhabiting them. Even in the middle of the Covid-19 pandemic, this initiative generated a lot of interest and became a social media movement.

Many of these efforts have a broader goal—to develop new ways of urban living that can help people and nature coexist in the city. Urban development plans view Mumbai's coasts as wastelands. In their minds, the only potential value of wetlands and coastal areas is as locations for real estate development, areas to be wrested from the sea. In contrast, the imagination of the people's collectives described here acknowledges the danger that the oceans will reclaim much of Mumbai's coastline in the decades to come. They envision the waterfront as a connected urban commons, one that fosters people's communities, while restoring coastal ecology.

In a city where most homes, offices and businesses are built just a few metres above sea level, flooding is inevitable, not just along the coastline and the creeks but also across the entire city. By 2070, Mumbai will have one of the highest populations in Asia exposed to coastal sea rise. Some estimates suggest that the entire financial capital region of downtown Mumbai may be underwater because of climate change. Rather than its current myopic obsession with economic growth fuelled by construction, this megacity needs to think about how to reclaim its once-symbiotic relationship with the ocean. Unless it does, the coasts and creeks of Mumbai will wrest back possession of the city—a testament to the futility of a grandiose colonial vision that attempted to create land out of water.

The Metro and the Well

Wells are especially important for Parsis, who use well water for religious ceremonies. One of the oldest Parsi wells in Mumbai, the Bhikha Behra Kua, is near Churchgate Railway Station. This well was built in 1725 by a Parsi merchant, Bhikhaji Behramji Pandey. Legend says that he was once arrested by the Marathas by mistake. When released, he built the well as a thanksgiving offering. The waters of this well are said to possess divine properties, granting wishes and healing the ill. Despite being so close to the sea, the well has sweet water. For over ten years, Parsis have conducted *humbandagi*, a community prayer, every month on Avan Roz (day of the water). This old well is now threatened by tunnelling for construction of the Mumbai metro. Another magnificent stepwell in the Dadar Parsi colony went dry for the first time in eighty-nine years in April 2017, possibly because of the construction of new buildings in the vicinity. This well provided water that was used in the daily rituals of the fire temple. The water for the rituals is now being replaced by tanker water. Local residents hope that the waters of the well will be replenished once the rainfall improves.

SIX

FANTASTIC BEASTS
AND WATER MONSTERS

In April 2019, a tweet from the Indian Army Twitter account said that an army mountaineering expedition had sighted footprints of the 'mythical beast Yeti' close to the Makalu Base Camp. They attached a photograph of the massive footprints, 32 in. long and 15 in. wide. The tweet attracted a lot of controversy on social media—not surprisingly, since social media is a place where controversy thrives. Whether the footprints belonged to an as-yet-undiscovered species, were a clever fake or something else altogether is something that only research can answer.

Cryptozoology is a branch of zoology that purports to deal with the study of reports such as this, that is about cryptids, hidden animals. Cryptozoology is defined as 'the study of animal species whose existence is not supported by empirical evidence, but rather hypothesized via indirect and uncertain information, including oral traditions, eyewitness accounts, and inconclusive physical evidence'. Many zoologists and biologists refuse to acknowledge this field of work as credible. After all, cryptozoology seems to encompass the study of creatures such as the Yeti, as well as creatures otherwise found in fantasy books and movies, like unicorns and fairies. But the International Society of Cryptozoology, which was active for a period of sixteen years from 1982 to 1998, and even published a peer-reviewed journal titled *Cryptozoology*, disagrees. The symbol of the Society is the okapi, a real species, initially described as an antelope with the markings of a zebra. The okapi was well known to people living in the area but acknowledged to be a different species by scientists only later. This legendary beast became the poster child of cryptozoologists. For them the okapi was proof that the careful observation and documentation of reports of mythical beasts could lead to discoveries of species new to science.

Many cryptozoologists recognize that only a few of the animals they document from folklore and reported sightings will turn out to be legitimate. But in between these two extremes—the careful research that leads to scientifically accepted reports of new species and the wild reports of sightings by charlatans churning out fakes—there is a vast grey area, where a lot of speculation exists but little is known.

It is no accident that many of these speculations are about water monsters. The purported Yeti footprints were found in the Himalayas. But in recent years, even the most inaccessible mountains have been so well explored that there seems little chance that a new species the size of a Yeti is hiding out in the Himalayas, still unknown to science (though one can never be sure). The waters of the oceans are far less explored though. What lies within their depths?

Bernard Heuvelmans, the twentieth-century French–Belgian zoologist and explorer, is considered the father of cryptozoology, coining the term from the Greek words *kryptos* (meaning 'hidden'), *zoon* (meaning 'animal') and logos (meaning 'study'). Heuvelmans was inspired by the Dutch zoologist Anthonie (Antoon) Cornelius Oudemans, who wrote the famous book *The Great Sea Serpent* (1892), which records 187 instances of reported sea serpent sightings. What should one do if one sees one of these beasts? The book gives detailed specifications.

> Voyagers and sportsmen conversant with photography are requested to take the instantaneous photograph of the animal: this alone will convince zoologists, while all their reports and pencil-drawings will be received with a shrug of the shoulders.
>
> As these animals are very shy, it is not advisable to approach them with a steamboat.

The only manner to kill one instantly will be
by means of explosive balls, or by harpoons loaden
with nitro-glycerine; but as it most probably will
sink, when dead, like most of the Pinnipeds, the
harpooning of it will probably be more successful.

European empires funded explorers who traversed
treacherous seas, seeking rich new lands to colonize.
Early maps of sea routes were filled with illustrations
of dragons and sea monsters. Many were marked
with the words *hic sunt dracones* (here be dragons)
to indicate dangerous, unknown territories. The crew
of the *Daedalus* claimed to have seen a snake-like
creature 18 m long, yellow-throated, with a horse-like
mane on 6 August 1848, off the coast of Namibia. In
Hook Island, Queensland, a report in 1964 described
a 24-m-long tadpole-shaped monster with dark bands,
resting in a lagoon. Cadborosaurus, reported from
Cadboro Bay in British Columbia, was described as
serpentine and horse-headed, with a serrated bony
ridge on its backbone. One of the most famous
mythical sea monsters is the kraken. With its waving
tentacles, the kraken is said to reach lengths of 15 m. It
is much feared by sailors as they were believed to drag
entire ships to the bottom of the ocean. Once restricted
to sailor's lore and fantasy books, the kraken became
well known across the world when it was featured in
the *Pirates of the Caribbean* movie trilogy. Zoologists
speculate that this mythical water beast could in fact

be the giant squid, or perhaps the even bigger colossal squid, which grows as large as 10 m.

Beyond the mighty oceans, many smaller water bodies—ponds, lakes, rivers and marshes—teem with accounts of strange creatures. Perhaps the most famous is the snake-like underwater monster believed to live in the Loch Ness Lake in Scotland. Popularly known as Nessie, she was first described in the 1930s. Underwater vehicles and sonar sweeps have fruitlessly explored the lake looking for Nessie. A 1975 scientific paper in the prestigious journal *Nature* even named her, giving her the scientific name of *Nessiteras rhombopteryx* (Ness monster with a diamond fin). Peter Scott, one of the authors of the paper, was a well-known naturalist who set up the Loch Ness Investigation Bureau in 1962 and later established the World Wide Fund for Nature.

Reports of extraordinary water creatures come from all continents. In Australia the bunyip, from aboriginal mythology, is believed to inhabit rivers, waterholes and swampy areas. Some say it looks like a giant starfish while others say it resembles a seal dog. In the African Congo River, legendary tales describe the mokele-mbembe, an amphibious, elephant-sized, long-necked and long-tailed herbivorous animal. Early-twentieth-century explorers searched assiduously for the beast but could not find it. A more recent expedition by a Danish crew in 2018 too failed to find any trace of the dinosaur-like animal though it did discover a new species of green algae, proof that there is much undiscovered biodiversity in the waters of the world, though it is more likely to be microscopic instead of giant-sized.

Many water bodies in India are also associated with folk tales and stories. *Amar Chitra Katha* comics talk of spirits and ghouls, celestial nymphs and water beasts who live in the depths of the waters. Of the Dashavatara (ten incarnations) of Lord Vishnu, the first incarnation is the *matsya* (fish) and the second is the *kurma* (tortoise).

An exhaustive list of water deities, sprites and mythical beasts described in diverse cultures, folk tales, scriptures and holy books would fill several pages. The *Little Mermaid* is a children's tale popularized further by Walt Disney Pictures, but the original folk tale may have been inspired by reports of mer-people in different cultures. The Afanc is a demonic creature from Welsh mythology that can take various forms, including that of a crocodile and a beaver, dragging its victims into the water to feast on them. Some say that the legendary King Arthur slew the Afanc, while others attribute the killing to his son Peredur.

Many lake monsters are described as serpents with long necks, but the Anfish from the marshes at the foot of the Tigris River in Iraq has hairy skin, the Balong Bidai of Southeast Asia is flat like a mat and the carbuncle of South American mythology has mirror-like shining gems on its head. Many of these lake monsters are believed to feast on people, attacking unwary boatmen and dragging them to the bottom of the lake. Some are believed to be venomous but others are said to bring good luck.

These beasts don't just look strange, they also have strange and unforgettable calls. The otter-like Ahuítzotl of Mexico is believed to make a sound like a baby crying, the Aidakhar of Kazakhstan has a trumpeting call and the Lukwata of Uganda bellows so loudly that the sound creates whirlpools in the water.

Cryptozoology may be a controversial, even a dubious science. But curiosity is not a dubious trait. Who among us as children did not want to encounter a fairy, a dragon or a water monster? We may have even conjured one up, speaking to it and making it part of our imagination. Perhaps this is why we nurture the cryptozoologist in ourselves.

The story of the duck-billed platypus, which nineteenth-century explorers dismissed as a hoax even after seeing its skinned pelt, is a cautionary tale for those too quick to dismiss wild tales of water beasts as imaginative folk tales. As Heuvelmans said, cryptozoologists must possess both patience and passion in abundance. While the discovery of new species is becoming rarer, history tells us they should not be dismissed without careful scientific study. For who knows what strange creatures we may discover in our backyard, or indeed our backwaters, like the microscopic algae discovered by the Danish crew that ventured up the Congo River to search for the mokele-mbembe.

The Lake of the Returned Sword

Sometimes the line between what is real and what is mythical can seem blurred. Until recently, the 1.6-km-long Hoàn Kiếm Lake (Lake of the Returned Sword) in downtown Hanoi contained a giant turtle. Many believed that the turtle was the same creature which received a magical sword from a fifteenth-century Vietnamese emperor. Cụ Rùa (which translates into Great-grandfather Turtle, although this turtle was female) was identified by biologists to be the Yangtze giant softshell turtle, an extremely rare species. The turtle died on 19 January 2016. Multiple wounds were seen on its flesh in the last few years, most likely due to pollution, garbage and sharp debris dumped in the lake. Huge crowds gathered to mourn her loss in Hanoi. The 170-kg turtle, considered by some biologists to be the last of her species, was later embalmed and placed on public display in a local temple.

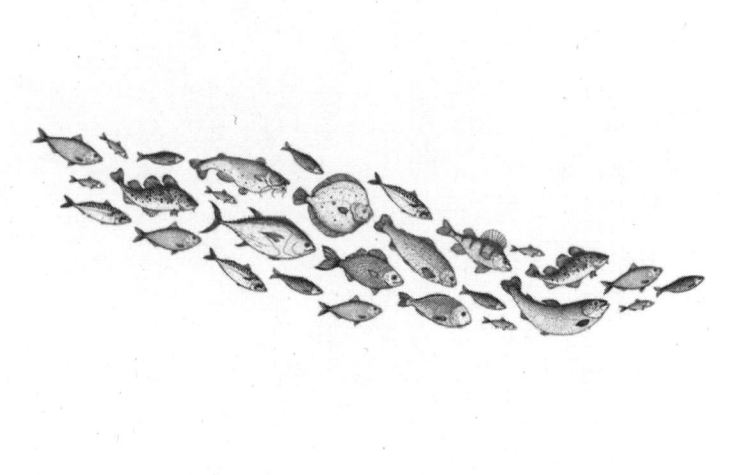

SEVEN

IS ALL WELL IN OUR CITIES?

Once-mighty rivers have been reduced to a trickle of sewage in cities. Lakes that used to stretch for acres are now a sea of concrete. What chance do tiny wells stand?

Those of us lucky enough to have looked deep into the waters of a well will remember the excitement of peering in to see our shimmering reflection, fringed by the trees in the background and framed by the sky. Of days spent with our friends in competition to cry ourselves hoarse, shouting into the deep shaft of a well to test out whose voice echoes the longest. And of squinting past the glare of the sun, trying to spot fishes, frogs, checkered keelbacks and Indian flapshell turtles lurking in the water.

India's cities have lost millions of wells in the past few decades. Heaped with garbage, filled in with concrete, covered by buildings—it is rare to spot a functioning well in the city. But if we look hard, we may be fortunate to spot some remnants of these water structures. Wells may no longer be a key feature of Indian cities today, but they provide a fascinating connection to the history and culture of cities and communities of the past, a connection we can still hope to regain and revive.

When large groups of people began to live together, they began to dig wells, seeking water that was clean and safe from pathogens. Rivers, streams and ponds were more likely to be contaminated by human and animal waste, agricultural residue or pathogenic microorganisms. Wells provided one of the safest ways to access clean, drinkable water from unpolluted sources deep underneath the ground.

Perhaps the oldest well discovered so far is a 5-m-deep tube well in the Mediterranean island of Cyprus. The well was dug between 9000 and 10,500 years ago, around 7000–8500 BCE, when humans began to build the first permanent settlements on the island. Other Stone Age wells built a thousand or so years later are still visible in the Jezreel Valley and the coastal areas of Atlit Yam in Israel. Several metres deep, these impressive structures were painstakingly dug from the ground, and lined with rough, undressed stones. The people who once lived here relied on wells for their supply of freshwater. When seawater from the coast intruded into the water table and filled the lower part of a well in Atlit Yam, the people filled the bottom of the well with large stones to reduce the seepage of saline water into the well.

Later Neolithic wells found in Europe used more sophisticated technologies of well construction, lining wells with wood to make them impervious to seepage and contamination. This period marked a transition in human lifestyles from nomadic to sedentary. Around this time the people of Central Europe, in the areas around Germany and the Czech Republic, began to live in one place with settled agriculture and livestock. For perhaps the first time they began to build permanent structures, using the wood from the forests around them. They also relied on wells to supply water. Once the earliest cities came up, wells became widespread. One of the most impressive ancient urban settlements is the Indus Valley Civilization city of Mohenjo-daro, built in 2600 BCE. Mohenjo-daro had over 700 circular, brick-lined wells—one well for every three houses!

As cities grew, more wells were dug in different parts of the world. Some were artesian, connecting to underground springs which filled the well with clean water. Others connected directly into the groundwater table. In mountain areas people dug a chain of wells along the slope, taking water from the topmost peak to the settlement in the valley via gravity. Wells could be dug directly into soft rock or clay soil. In arid or semi-arid soil, where a well was likely to collapse, it was lined with wood, bricks, terracotta rings or rough/

dressed stone, sometimes sealed by lime mortar. Today of course, most wells in India and many other parts of the world are lined with cement rings.

Chanakya's third-century administrative treatise *Arthashastra* provides insights into the importance of wells in ancient times. According to the *Arthashastra*, a city must have wells in different locations, on the parapet of forts, in private homes, even in prisons. Chanakya also specified the kinds of crops to be grown around wells, mostly vegetables and edible roots. The monarch needed to take on the responsibility of constructing wells in areas that lacked water. Wells were important, requiring protection. No one could dig a well on another's land and anyone found polluting a well would receive punishment.

A variety of wells can be found across India today. They are called by different names across the country. The names indicate not just the type of well but also the nature of ownership. In Rajasthan, a private well is a *kuan* but a community well is a *kohar*. Stepwells are larger wells extending below the ground, with steps along the walls that go down to the base. As the water level rises in the monsoon and falls in the summer, the steps allow people to walk down to the level of water, filling their pots with ease. A 5000-year-old stepwell in the ancient city of Dholavira during the times of the Indus Valley Civilization is 73.4 m long and 29.3 m wide, three times as large as the famous Great Bath of Mohenjo-daro.

With their perfect symmetry and geometrical pattern, stepwells are a fascinating form of architecture that is both aesthetic and functional. The size and shape of the well depends on the type of soil and the depth of water. Many stepwells have a single flight of steps but some have multiple sets of stairs, often arranged in a fractal or self-similar pattern when viewed from the top, with their staircases and pavilions repeating across levels. More recently, research shows that fractal architecture can bring down stress levels, making some architects speculate that the artisans who built these wells knew the importance of fractal designs for stress relief.

Stepwells are an important source of water in some of the driest regions of the country. In Rajasthan they are called baolis or *jhalara*s, and in Gujarat they are referred to as *vav, vapika* or *vavadi*. Some large stepwells were built on travel routes between settlements and served as a place of rest for travellers. Others were built inside towns and villages and used for social gatherings. The vavs of Gujarat have exquisite architecture that is a treat for the eyes. They range in design and complexity, from a simple flight of steps leading down a single entrance, to elaborate structures with as many as four entrances, and as many as seven sets of stairs with carved pillars, columns and pavilions leading down to the water. The walls and pillars are often covered with rich carvings of gods and goddesses, elephants, birds, horses and floral designs. Some older stepwells also have Buddhist commandments inscribed on the walls though the walls of the later stepwells, built by the Mughals, are often plain.

Stepwells also have shrines dedicated to family goddesses and local deities. In Saurashtra, a local legend describes a well that was dry for years being revived after a newly-wed couple descended into it, sacrificing themselves as an offering to the goddess. It is said that as the couple descended deeper into the well, the water began to flow, drowning them but saving the village. The construction of a stepwell was believed to be an act of philanthropy, one that helped the village and also bestowed blessings on the patron. The stepwells of Gujarat are unique. Many of them are believed to have been built by women from diverse sections of society, rich and poor, queens and servant girls. One of the most magnificent stepwells of India is the Rani ki Vav, built in 1032 CE by Queen Udayamati. The well, found in Patan at a short distance from the city of Gandhinagar, is now recognized as a UNESCO World Heritage Site. The walls of the well are adorned with sculptures of women engaged in daily tasks such as combing their hair or stepping out of a bath, portraying them as being engaged in a range of relationships as friend, lover, companion and mother.

The exquisite beauty and grandeur of Rani ki Vav and other equally magnificent stepwells such as the horseshoe-shaped stepwell of Chandrasekarapuram in Prakasam (Andhra Pradesh), the helical spiral stepwell of Champaner and the Adalaj Vav near Ahmedabad (Gujarat), the fractal triangular stepwell of Abhaneri (Rajasthan), the Pushkarinis of Hampi (Karnataka) and the Lolarka Kund of Varanasi (Uttar Pradesh), leave us in awe. But this awe is tinged with sorrow— such skill and workmanship can no longer be found.

Not many residents of Delhi know that the city once had several stepwells, of which at least thirty-two still remain. While people visit the city to see the popular monuments like the Qutub Minar or Humayun's Tomb, the stepwells, gems of architecture, remain hidden from tourists and locals. The rectangular Agrasen ki Baoli, believed to have been built by Raja Agrasen in the tenth century CE, is better known because of its location near Connaught Place, the commercial hub of New Delhi. But many other stepwells in Delhi are worth visiting including the circular Hazrat Nizamuddin Baoli near the *dargah* (shrine/tomb) of the Sufi saint Hazrat Nizamuddin Auliya, and the round baoli in Feroz Shah Kotla Fort.

Stepwells in south India include those built by the Vijayanagara kings and the Adil Shahi rulers. The *bavadi*s or baolis of Bijapur (now Vijayapura) are large square structures of different sizes. Two of the largest, Taj Bavadi and Chand Bavadi, were built as memorials to the queens. At one point in 1819 Bijapur had 700 stepwells within the fort walls, enabling the city to withstand long sieges. Today, many of these bavadis lie in a dilapidated state, dry, full of polluted water or garbage, and even used as urinals. Fortunately, a new initiative by the Vijayapura City Corporation has cleaned and restored many of these iconic stepwells including the Taj Bavadi and the Chand Bavadi, with the eventual goal of using these to provide water to the city once again.

In Bengaluru, open wells were a common feature once. In a map of 1885, 1960 wells were marked in the centre of the city, providing water for homes and communities. In 2014, when we went looking for these wells, we could only find forty-nine. We did however come across a few historic wells and remnant structures. In the periphery of the city, we found a small but beautifully constructed well inside the walls of the mud fort of Begur, Bengaluru's oldest settlement with a documented history dating back to 890 CE. Another well in the city centre near the bustling Corporation Circle, was so large that it had nineteen pulleys. While that well no longer exists, we were thrilled to come across a similar set of pulleys on a nearby dilapidated well.

Few of Bengaluru's wells now remain. Most are polluted, crumbling and no longer in use. They lost their importance when piped water began to be provided to the city in the 1890s. Wells, once considered sacred, began to be used as places to hide the bodies of those who died during the plague outbreak of the 1890s. As the city expanded and agriculture ceased to be practised, wells further lost their significance. Many wells fell into disuse, being filled in and built over.

Thankfully some cities like Vijayapura have begun to revive their wells. When Jodhpur in Rajasthan experienced a severe drought in 1985, citizens came together to clean up the Tapi Bavadi. The revived bavadi began to provide water during the drought. But urban memory is fickle and short-lived. After a heavy monsoon in 1989 the local water supply improved, and the bavadi was once again neglected. In recent years, thanks to the efforts of an elderly Irishman Caron Rawnsley, along with the Mehrangarh Museum Trust and the Indian National Trust for Art and Cultural Heritage (INTACH), a number of stepwells have been cleaned up and restored once again. The Tapi Bavadi is one of these. Rawnsley is the grandson of the British chief railway engineer who built the railway line from Mumbai to Vadodara. Many local residents called him *paagal saab* but a number of people have now joined him, forming a larger movement. The stepwells they restored support a variety of animals including bats, birds, fish, turtles and snakes. Rawnsley takes groups of local residents and schoolchildren to the stepwells, teaching them about their local history. He says that the stepwells are essential for Jodhpur's future survival in a time of climate change, when droughts will become increasingly frequent.

In Bengaluru the Biome Environmental Trust, spearheaded by water conservationist 'Zenrainman' S. Vishwanath has initiated a Million Wells campaign. Working with other groups, Biome has revived wells in private homes, apartments, temples, schools, parks, *dhobi ghat*s (place for washermen to wash clothes) and even in a police station. They work with *mannu vaddar*s, traditional well diggers who possess a wealth of knowledge about how to dig and restore wells. Vishwanath moderates a Facebook page, Open Wells of India and the World, where over 5000 members post photographs and describe the beautiful wells they discover on journeys across the country.

The experiences of Vijayapura, Jodhpur and Bengaluru bring us hope that the wells in our cities, once lost, can be rediscovered and revived. Along with water, lost urban heritage is revived. When wells are restored, children will once again experience the joy of looking into the shimmering waters of a well, shouting into its watery depths, waiting to hear the echoing response.

Neolithic Carpenters, Builders of Wells

In 2018, while building a large highway in the Czech Republic, people stumbled across the remnants of a wood-lined well in the tiny town of Ostrov. The well is 4.5 ft deep, with a square base, and lined with oak wood. Because the wood was surprisingly well preserved, archaeologists were able to examine the tree rings in the oak wood to estimate the age of the trees. They corroborated the age by radiocarbon dating charcoal fragments found in the well. Their findings indicate that the trees used to construct the well were cut down around 5255 BCE, making the well 7200 years old. The well was built at the end of the Neolithic period, during the New Stone Age. The structure of the wooden lining, which was built by inserting planks of wood into grooved posts at the four corners of the base, is remarkable, considering that these carpenters worked with crude tools made of stone, bone, wood and horns.

More recently, archaeologists from Germany examined the wood from four ancient wells, dating them to between 5469 and 5098 BCE. They concluded that the Neolithic farmers displayed 'unexpectedly refined carpentry skills'. Lacking metal, they used fire, burning charcoal and wooden adzes or axes to cut and shape the wood. During the time these wells were built, other evidence shows that the area went through a series of droughts and floods. Wells such as the one in Ostrov, built by farmers who doubled up as carpenters, played an important role in providing water security during unsettled times.

EIGHT

KOLKATA: TRANSFORMING WASTE TO WEALTH

Every city has its iconic spots such as the Gateway of India in Mumbai, India Gate in Delhi and Marina Beach in Chennai. For Kolkata, surely it is the Howrah Bridge. Strung across a river like a gigantic spider's web, this suspended cantilever bridge has been around for decades. It was inaugurated in February 1943 without much fanfare—the Second World War was in progress and the British did not want to attract the attention of the Japanese. Unlike earlier constructions of that time where the raw materials were usually brought from Europe, more than 90 per cent of the raw material used to build the Howrah Bridge came from India. A true 'Made in India' product!

Passengers tend to ignore the muddy river below the Howrah Bridge. Yet this river plays a far more important role in the history of the city. Calcutta, as the city was known in British times, was the first colonial capital of India—before the capital moved to New Delhi. In 1690, Job Charnock of the British East India Company chose the village of Sutanuti, on the banks of the Hooghly, as the British headquarters in Bengal. This served as a strategic military location for the British—at a safe distance from Mughal strongholds, it was protected by the Hooghly on the west and the saltwater marshes on the east. Its riverine location also provided convenient access to the sea, helping the British to construct a port that they could use for shipping and trade.

From the very beginning, the British considered Calcutta a city of disease, calling the eastern marshes and saltwater lakes 'rank vegetation, producing constant and unwholesome exhalations'. Captain Alexander Hamilton, who travelled to India in the eighteenth century, wrote that Job Charnock:

> Could not have chosen a more unhealthy place on all the river; for three miles to the north-eastward, is a salt-water lake that overflows in September and October, and then prodigious number of fish resort thither, but in November and December when the floods dissipated, those fishes are left dry, and with their putrefaction affect the air with thick stinking vapours which the north-east winds bring with them to Fort William, that they cause a yearly mortality.

By the 1770s, thousands of people had died in epidemics of cholera and fevers, compounded by famine. Concerns about the drainage and sanitation of Calcutta became even stronger. The city was growing in size and importance. British planners wanted to ensure that the city was free of disease and death.

In order to do so, they needed to figure out how to dispose of the city's waste. The Hooghly seemed like a natural solution. But the topography of Calcutta was against them. The river lay on the western side of Calcutta, but the natural slope of the land was towards the east. Sewage was sent into the river, but only some of it went into the Hooghly. The rest of the sewage tended to flow east towards the marshy saltwater lakes. During the monsoon, heavy tidal action swept the sewage that did enter the Hooghly back into the canals.

The Governor General, Lord Wellesley, wrote in 1803: 'An original error has been committed in draining the town towards the river Hooghly; and it is believed that the level of the country inclines towards the salt water lake, and consequently that the principal channel of the public drains and water-courses ought to be conducted in that direction.'

Between the 1820s and 1850s, several schemes were proposed to improve the drainage of the city's waste. In 1821, Lieutenant Schalch suggested that sewage could be taken from the city into the lakes. He also suggested that the sewage-filled lakes could be used for cultivation by raising the level of the marshy land, and clearing vegetation from this area, making the lakes economically productive. Captain Prinsep recommended the transfer of silt from the Hooghly to fill in the saltwater lakes, further expanding the cultivable area. F.P. Strong, a British surgeon, suggested that the saltwater lakes could be drained and the marsh used to grow commercial crops like paddy and indigo. By the early nineteenth century the saltwater lakes became a profitable source of income for local zamindars who profited from activities such as fishing and agriculture.

But what of the disposal of waste? A scheme proposed by W. Clark, Secretary to the Municipal Commissioners, was eventually approved. This scheme relied on the natural topography of the city, with its gradual slope towards the east. A sewer system was designed to collect a mix of different kinds of water—subsoil water from the marshy parts of the city, kitchen waste and sewage from homes, waste from animal stables, and rainwater—draining into the saltwater lakes. Alongside, the plan was to reclaim the marshy lands and saltwater lakes for cultivation by filling them in. David Smith, the Sanitary Commissioner for Bengal, approved of this plan, saying in his *Report on the Drainage and Conservancy of Calcutta* (1869) that this would ensure the wetlands would 'speedily become one of the most fertile districts in Bengal'.

In the meantime, waste from Calcutta was already being transported by the municipal railway line and dumped in a square patch of land in Dhapa (the Dhapa square mile) in the saltwater lakes as part of an experiment at reclamation. After decades of planning by numerous committees, recorded in consultations, reports, plans and schemes, the grand plan for treating Calcutta's waste was put into place at the end of the nineteenth century. A network of storm water and sewage canals was created, carrying rainwater and sewage from Calcutta to the saltwater lakes. These once-pestilential marshes began to provide two extraordinarily critical services to the city—cleaning its waste and growing its food (through agriculture and fishing).

But as the ever-insatiable city continued to grow, it needed new land on which to expand. Calcutta cannibalized itself, casting its predatory gaze on the saltwater lakes that were so important for its health and survival. As early as 1933, British reports began to speak of 'further raising and filling in of the Salt lakes to make room for future expansion of the city'. By 1951, Calcutta was congested, with a population of 2.5 million, and 140 persons per sq. km. Hemmed in by the Hooghly River on the west, the city could only expand into the saltwater lakes on the eastern side. In 1967, a 10.3 sq. km area in the north-western part of the saltwater lakes was transformed into the township of Salt Lake City, reclaimed using silt from the Hooghly.

Although diminished in size, cultivation and fishing continued in the wetlands. In the 1960s, the area around the Dhapa square mile, the initial experimental site that the British reclaimed for cultivation, supplied nearly 50 per cent of the city's vegetables. *Bheri*s, ponds of varying size containing a mix of sewage and freshwater, were used to grow fish across an area covering around 260 sq. km. Meanwhile, the city continued to grow, continuing to grab land from the salt lakes. From an area of 182 sq. km in the nineteenth century, these lakes had shrunk to 83 sq. km by the 1950s. By the 1990s, the Salt Lake City township, the Eastern Metropolitan Bypass, the Anandapur Kasba township, the New Town Rajarhat and other urban development had further impacted the biodiversity of the wetlands, and the livelihoods of local farmers and fishers.

Several committed individuals and civil society organizations attempted to protect the wetlands. Most prominent among these was Dhrubajyoti Ghosh, a sanitation engineer with the Government of West Bengal, who called the wetlands the 'kidneys of Kolkata'. Ghosh's efforts helped protect the wetlands. In 2002, 125 sq. km of the East Kolkata Wetlands (EKW) were declared as a Ramsar site, protected under the international Ramsar convention on wetlands.

Outside the boundary of the Ramsar site, the growth of the city continues to eat away at the wetlands. Even within the Ramsar site, the wetlands are not completely protected. But even as new roads, homes and business complexes come up across the

EKW, many aspects of the old terrain are still visible, revealing a fascinating mosaic of land and water. The blue of the bheris and the green of the fields are interspersed with small homes, kitchen gardens and ponds of varying sizes. Fields of rice are cultivated alongside neat rows of seasonal vegetables using a mix of freshwater and sewage. Bheris are of different sizes, owned by individuals, private groups or cooperatives.

The right blend of freshwater and sewage needs to be maintained to foster crop productivity and fish growth, while ensuring the produce is not affected by disease. Farmers and fishers rely on traditional ecological knowledge, passed on through generations. Fishers need to know the species of fish that thrive at specific depths of water, and to understand the tolerance of different fish species to sewage. This knowledge is passed down from father to son, modified to changing circumstances. Rice husk, cow dung and mustard peels were added as fish food to the water earlier. Fishers now include meat waste from butcher shops that have come up in the neighbourhood, as the city expands into the wetlands.

The siltation of the canals and the growth of weeds have reduced the inflow of water over the years. The composition of waste has also changed. Once organic and non-toxic, Kolkata's waste now contains toxic chemicals and industrial effluents. Household waste has also changed in composition, containing non-biodegradable materials like plastic. This impacts fish health. Fishers say they need to spend increasing sums of money to purchase medicines to treat fish infections. They supplement their reliance on medicines with prayers. During the monsoon, devotees of all faiths visit the dargah of Pir Mobarak Gazi at Ghutiari Shariff, returning with saffron flags sprinkled with holy water. The flags are tied to poles and placed in the bheri to protect it from harm.

The wetlands hold sacred, spiritual and cultural meaning for local residents, a value which extends well beyond the ecological and monetary. In addition to bheris, the landscape is dotted with *pukur*s (small ponds). These are community spaces where men and women, old and young, gather, sometimes talking for hours. Some pukurs are used for bathing. Others located next to temples are considered sacred, with cemented sides and steps. Meat wastes are not added to sacred ponds, as they believe this would defile them. Pukurs have begun to disappear, giving way to buildings. As they vanish, the community culture and social capital that once characterized these areas has begun to give way to a more fragmentary, disconnected form of urban living where people do not know their neighbours.

The wetlands contribute to the informal economy of the city, also providing invaluable ecological services through the treatment and removal of waste. Rice, vegetables and fish cultivated in the wetlands make their way to the markets in Kolkata, and from there to the dinner tables of homes across the city. The wetlands create social ties and foster the transfer of traditional social-ecological knowledge of farming and fishing through generations.

In large cities like Kolkata, less than 50 per cent of the city's effluents are connected to a sewage system. In smaller towns in India the situation is even worse. Almost 85 per cent of the sewage is untreated, directly dumped into water bodies, converting urban rivers and lakes into cesspools of human and animal waste.

Some people consider the challenge of urban waste to be an unsolvable problem—an inevitable consequence of a developing economy like India, where everything is cheap, easily purchased and just as easily discarded. But who disposes of the waste? In India, waste disposal is almost always connected to caste exploitation. Across the country, in small towns and large cities, Dalit waste pickers, sanitation workers and manual scavengers work in exploitative conditions, are poorly paid and lack protective gear. Yet as the case of Kolkata shows us, waste management can also be a source of wealth, providing a way to live with dignity.

India has an advantage when compared to many other countries. Close to half of the waste that cities generate is organic. More cities could follow the path of Kolkata, using their wetlands and water bodies to clean the waste, turning garbage into manure, and reaping the benefits. Sadly, Kolkata's unique history of turning liquid faecal waste into hard cash is in danger of being forgotten even within the city. Few residents know that the fish and vegetables they eat are fed by sewage they have produced, a true circular economy. India's cities have much to learn from Kolkata's example by protecting their wetlands, ensuring that they in turn help the city, by turning waste into wealth.

NINE

TINKER, TAILOR, MAPPER, SPY: SECRET EXPEDITIONS TO MAP RIVERS

Colonial empires were fanatically keen to survey, map and record natural resources in their rich colonies. The British initiated the Great Trigonometrical Survey, a colossal mapping exercise covering the length and breadth of the Indian subcontinent between 1802 and 1883. The vast Indian subcontinent held many secrets. One mystery that plagued explorers for decades was that of the origins of the Brahmaputra River. The first map of the mighty river was prepared in 1733 by two lamas under the supervision of the French geographer Jean-Baptiste Bourguignon d'Anville. This map depicted the Tsangpo of Tibet and the Brahmaputra of India as being two different rivers. The pioneering English geographer and surveyor James Rennell surveyed the Indian side of the landscape between 1763 and 1782. He was of a different opinion, concluding that the two rivers were the same. Surgeon Major L.A. Waddell, writing in 1868, agreed with Rennell, pointing out that in Tibetan 'Tsang-pu' meant 'son of Brahma', i.e., Brahmaputra. Yet unless they mapped the river on the Tibetan side, they could not be sure.

This seemed impossible. The Chinese emperor, who controlled Tibet and Nepal in the nineteenth century, closed the borders between Tibet and India in the 1850s. The Great Game (a term popularized by Rudyard Kipling in his novel *Kim*) was then at its peak, with the British and Russian empires fighting for control over the territories of Central and South Asia. It was not just the Brahmaputra that the British wanted to map—they were desperately keen to collect intelligence on the entire region of Tibet. But Europeans, with their pale skin, stood out as foreigners at the border. Turned back from the border, sometimes killed for trying to cross over without permission, they could not map the river systems of Tibet.

Thomas G. Montgomerie of the Survey of India had a brainwave. Caravans of merchants traversed the border, as did cattle herders and lamas. Why not train Indians in the science of mapping and send them across the border as spies? His first attempt ended in tragedy. Abdul Hamid, Montgomerie's first recruit, made it to

Yarkand (now part of Turkestan) in 1863 but died on his way back while navigating the Karakoram mountains (Montgomerie later recovered and used his notes).

The British turned their attention to the Bhotias, a mountain community of Tibetan origin living in the Kumaon and Garhwal valleys who could pass off as Tibetan with ease. Many of them were also familiar with the routes into and out of Tibet, since they traded grain with the Tibetans for other materials such as salt, wool and ponies. Many traders held unremarkable daily jobs as tailors and teachers, given the honorific title of 'Pundit'. Montgomerie recruited two cousins, Nain Singh and Mani Singh, for his next expedition. Nain Singh was an English-speaking schoolmaster, and his cousin Mani Singh was a *patwari* (local revenue official). The two young men had previously accompanied the Schlagintweit brothers, Germans who conducted surveys for the East India Company in the 1850s. Yet they needed further instruction.

Training the cousins in the art of surveying took many years and was conducted at the headquarters of the Survey of India, at Dehradun. Surveyors in India carried thermometers, graduated ropes, tape and other survey gear with them. But the surveys in Tibet had to be done in secret. If the men were caught, they could be jailed, even killed.

The terrain was hilly, and the size of the steps they took varied depending on whether they were climbing uphill, or downhill. The cousins were taught to take standardized steps of 33 in. in length, irrespective of the terrain, so that they could convert the number of steps into a standardized measure of distance—a habit that took some effort to learn. Disguised as monks, they carried prayer beads in their hands. In contrast to the Buddhist string which had 108 prayer beads, theirs had been modified to hold a hundred beads. Every time they completed a hundred steps of a standard length, they moved to a different bead, thus keeping track of the distance travelled. As Montgomerie explains in his *Report of a Route-Survey Made by Pundit, from Nepal to Lhasa, and Thence Through the Upper Valley of the Brahmaputra to Its Source*:

> It was necessary that the Pundit should be able to take his compass bearings unobserved, and also that, when counting his paces he should not be interrupted by having to answer questions . . . Whenever people did come up to the Pundit, the sight of his prayer wheel was generally sufficient to prevent them from addressing him. When he saw anyone approaching him, he began to whirl his prayer-wheel round and as all good Buddhists whilst doing this are supposed to be absorbed in religious contemplation he was seldom interrupted.

Sextants were used to record latitudes, and the mercury required for the sextant was hidden in coconuts.

Additional mercury was concealed in cowrie shells sealed with wax. At night, when the other members of their travel party slept, the Pundits made their way to a hidden spot, lighting a fire. Using the mercury thermometer that they kept hidden in a prayer bowl, they recorded the temperature at which water boiled. From this, they calculated the air pressure, and estimated the elevation. They recorded land routes, prices of goods and other information that might be useful to the British, writing detailed notes that they rolled up and secreted inside their prayer wheels.

In the true manner of spies, each of them was given a code name. Nain Singh was known as 'The First Pundit', 'The Chief Pundit' and 'The Pundit', while later recruits Kishen Singh and Nem Singh were known as A-K and N.E.M., respectively. The cousins left Dehradun in January 1865, reaching Kathmandu in March. After a few unsuccessful attempts to cross over into Tibet, they separated to try different routes. Mani Singh returned to Dehradun, unsuccessful. Nain Singh was more persistent. He disguised himself as a Ladakhi, trying to cross over into Nepal. Despite being swindled of his money, he managed to join a caravan entering Nepal. Once there, he feigned illness, separating himself from the group and making his way to Lhasa in Tibet. On his way, he witnessed three men drowning in front of him in a tragic accident on the Tsangpo.

Nain Singh reached Lhasa on 10 January 1866 and stayed there for several months but was unable to access the Brahmaputra as it was too dangerous. He walked back to Dehradun, returning home after a two-year journey. The detailed notes he carried enabled the Great Trigonometrical Survey to map the route of the Tsangpo from its origins near Manasarovar Lake in Nepal up to Lhasa in Tibet.

Nain Singh went on other missions to survey the headwaters of the Sutlej and Indus rivers in 1867, but it was in his fourth and last mission under Captain Henry Trotter, in 1873, that he once again crossed the Tsangpo, surveying 48 km of the river for the first time, and recording its approximate route for another 160 km. This came at a huge personal cost. The harsh sun and climate had affected his eyes, ageing him prematurely. His days as an explorer had ended, but he continued to help with training new recruits until his death in 1895.

Nain Singh's cousin Kishen Singh took on additional surveys, attaining even greater fame. On his fourth and final mission in April 1878, Kishen Singh and his servant Chumbal Singh set out from Darjeeling, on what would become an epic four-year journey. They intended to go northwards via Lhasa to Mongolia and try to enter China. They were robbed of everything by a band of thieves. Chumbal lost his toes to frostbite, and another servant Ganga Ram deserted them, taking their money, horses and guns. While taking a circuitous route back to India, which involved a detour via Lhasa, Kishen Singh found that the Tsangpo flowed farther westwards than was previously thought, giving more credence to the possibility that the Tsangpo was the Brahmaputra of India, and not the Irrawaddy River of Burma.

Where could the Tsangpo connect with the Brahmaputra? Was it through the Dihang or the Subansri tributaries? From the Indian side, surveyors had only been able to map the Brahmaputra up to a location called Sadiya (located in the present-day Tinsukia district of Assam). The upper reaches of the river were inhabited by the Abors, a tribe that was hostile to outsiders. Even the brave Pundits could not be convinced to map these territories, covered with thick jungle, full of wild animals and guarded by the Abors. The surveyors decided to map the river southwards from the southernmost point that Nain Singh reached in 1874, Chetang in Tibet.

They recruited a lama, Nem Singh, who took along an assistant Kinthup, a tailor from Darjeeling. They reached Lhasa in October 1878 and made their way 466 km east along the Tsangpo to a village called Gya La Dzong, where they found the river entering a deep, impassable gorge. There were just 257 km of uncharted river territory, but how could they hope to map it? They returned to India, leaving the mystery of the connection between the Brahmaputra and the Tsangpo unsolved.

In July 1880, Lieutenant Harman of the Great Trigonometrical Survey sent a Chinese lama, accompanied by Kinthup, asking them to make another attempt to map the stretch of the Tsangpo from Gya La Dzong to the plains of India. If that was not possible, he instructed them to throw logs into the river at the lowest point they could access. Harman set watchers on the Indian side, at Dihang, to look for these logs. If they found them, they would know that the two points were connected, confirming the route by which the Tsangpo became the Brahmaputra.

Kinthup and the Chinese lama took a long time to reach Gya La Dzong. The lama had an affair with the wife of a host in another village and was held captive for four months. This incident was a portent of things to come. He later disappeared, after selling Kinthup as a slave to a fort master. Kinthup escaped after nine months, continuing on his mission alone. He was chased by the fort master and sought refuge in a monastery. The chief lama at the monastery secured his release but then claimed him as his servant. After four months of hard labour, the lama allowed Kinthup to take leave, supposedly to go on a pilgrimage.

Kinthup was a man of fortitude. He could have hotfooted it back home but decided to continue with his mission to map the river. He cut 500 logs and hid them in a cave, and went to Lhasa, where he sent a message to Lieutenant Harman to inform him that he would throw fifty logs every day into the river. Unfortunately, the message never reached Lieutenant

Harman, who had waited two years for Kinthup but left India after he fell sick.

Kinthup returned to the monastery and served the lama for another nine months until he was released. He then made his way to the cave, retrieved the logs and threw them into the Tsangpo as promised, eventually returning to Darjeeling in November 1884, more than four years after he had left. But his return was never celebrated, nor his efforts recognized. Even worse, two years after his return, when he narrated the details of his journey from memory, his report was not considered authentic. Lieutenant Colonel G. Strahan, officiating deputy surveyor-general, Trigonometrical Branch, said in a report: 'This man not being a trained explorer, the information he brought is not based on a route survey, and can only be regarded as a bona fide story of his travels, related from recollection two years after his return.'

By the early twentieth century, after a few additional expeditions by the British, the last pieces of the puzzle had been filled in. The Pundits and other Indian explorers played a critical part in this effort, but what was it that motivated the Pundits to undertake these perilous journeys? For some, it was the opportunity of better employment. They received salaries ranging from Rs 16 to Rs 20 per month, which constituted a large sum for them at the time, though small by British standards. For others like Kinthup, it was an overdeveloped sense of duty, even when they received little monetary benefit.

Meanwhile, what recognition did the Pundits get for their work? Nain Singh, the Chief Pundit, was honoured by the Royal Geographical Society in 1868 for his route survey from Manasarovar Lake to Tibet, receiving a gold chronometer worth at least £30 then. But even here he shared credit with his English employer.

Sir Roderick Murchison, in his address to the Royal Geographical Society on 25 May 1868, said:

> That the Pundit, while maintaining his disguise, should have been able, amid a watchful and suspicious people, to keep upon so long a line a careful road-book with a full record of bearing and distances, and a very extensive register of observations, is certainly no ordinary feat; and reflects infinite credit, not only on the individual employed, but on Captain Montgomerie's judgment in selecting him for the duty.

In 1877, Nain Singh was also awarded the Victoria or Patron's Medal. Colonel H. Yule, who received the medal on his behalf, praised Nain Singh saying,

> He is not a topographical automaton, or merely one of a great multitude of native employes with an average qualification. His observations have added a larger amount of important knowledge to the map of Asia than those of any other living man, and his

journals form an exceedingly interesting book of travels.

He received a land grant from the British government and was featured on an Indian postal stamp in 2004.

Kishen Singh also received a medal from the Paris Geographical Society and a monetary award from the Royal Geographical Society. He was conferred the title of 'Rai Bahadur' by the British government in India, granted a revenue-generating village and lived well till his death in 1921 at the age of seventy-one.

Kinthup, the poor tailor from Darjeeling who perhaps endured the most hardship but mapped a crucial part of the river, received neither accolades nor medals, though he did receive a large gratuity of Rs 3000 from the Royal Geographical Society. We do not know what happened to the money, only that he returned to his old life as an impoverished tailor in Darjeeling.

In these explorations, one name was almost erased from history—Chumbal, cook, porter and faithful companion to many of the Pundits including Nain and Kishen Singh. Chumbal went through the same hardships that the Pundits did, falling sick, losing his toes to frostbite and suffering snow blindness while staying true to both his masters and the mission. But there seems to be no record of what happened to him.

Written by the British, the history of the Great Trigonometrical Survey fails to do justice to the Indian tinkers and tailors who went on to spy for the

British in treacherous foreign lands. Even when they acknowledged the role of the Pundits, they saw their role as limited merely to the gathering of information, with the English surveyors playing the important role of synthesis. They did not teach the Pundits how to make maps, or to use the data they collected for analysis. In recent years, efforts are being made by Indian researchers to remedy this historic injustice. In their 2006 book *Asia ke Peeth Par: Pandit Nain Singh Rawat*, professors Shekhar Pathak and Uma Bhatt use Nain Singh's diaries, written in *khadi boli* (an old form of the Hindi language), along with records of his journey retrieved from the archives in London and Dehradun, to provide a fuller account of his intrepid journeys.

The Europeans used the term '*terra incognito*' to describe many of the riverine mountainous regions they mapped, overlooking the fact that many of these landscapes were not incognito to the Indians who resided there and traversed these routes. British surveyors relied on a retinue of Indians who acted as guides, cooks, attendants, cleaners and cart drivers. Meanwhile, the Indian spy became a dispensable casualty of the surveys, occasionally celebrated and rarely mourned. Perhaps we will never know about many such explorers who perished in the colonial explorations, the adventures they had and the hardships they faced. But there may be many like Nain Singh whose stories lie in dusty archives, waiting to be retrieved, rediscovered and shared with the world.

TEN

SHACKLING THE WATERS: DAMS AND CITIES

Many of India's smaller cities, with populations between 50,000 and 1,00,000, receive about half their water supply from groundwater. The rest of their water comes from nearby rivers and lakes. In metropolitan cities, groundwater use drops to about 12 per cent. Large cities, with millions of people, and intensive residential, commercial and industrial requirements for water, cannot get their water from local sources alone. They need to supplement their local water supply by piping in water from distant sources.

Bengaluru, a city of 12 million people, once received most of its water supply from local lakes and wells. The city now receives water from the Krishna Raja Sagara Dam on the Cauvery River. This water is pumped to the city over a distance of 100 km and up a height of over 500 m, consuming substantial amounts of electricity and generating a high fossil fuel footprint.

A sentimental story by the famed Bengaluru writer D.V. Gundappa described an incident at the funeral of Sir Mirza Ismail, former Dewan of Mysore (now Mysuru), in January 1959. People had gathered to mourn his death. An elderly lady seemed especially devastated, sharing her memories of the Dewan. When she was younger, and pregnant, Sir Mirza had noticed her carrying water from a public tap to her home in Bangalore (now Bengaluru), a large burden for a woman in her condition to carry on a regular basis. Within a couple of days, he ensured a public connection was provided right next to her home.

Taps are certainly more convenient than hauling water from a well or taking pots to the river, especially for women, who are usually stuck with the task. But to function well, taps require a reliable supply of water. Across India, cities get water from local rivers and lakes. When this supply runs out, because of increasing population and consumption, the city turns to more distant sources, building dams across rivers to create reservoirs, and pumping it to the city for processing and distribution. Dams serve a range of needs: flood control, hydropower generation, and water supply for irrigation, industrial and domestic use. The National Register of Large Dams 2019, compiled by the Central Water Commission, notes that India has 5334 completed large dams, with 411 more under construction. Spread across the country, water-guzzling cities depend on these dams to supply their expanding needs. Thus the Hirakud Dam on the Mahanadi supplies water to Sambalpur city, and the Tehri Dam brings water to Delhi across a distance of nearly 300 km.

Ancient cities relied on smaller reservoirs and check dams. In 2015, scientists working on the archaeological site of Dholavira in Gujarat used ground-penetrating radar to locate a set of small, interconnected reservoirs buried below the ground. The location and structure indicated that this Indus Valley city, which flourished between 3000 and 1700 BCE, had developed a sophisticated system to deal with floods. The interconnected set of reservoirs slowed down the flow of water from these tanks into a larger reservoir on the eastern side, protecting the city from flash floods.

In the second century CE, the south Indian Chola King Karikala commissioned the construction of the Grand Anicut to control flooding on the Cauvery River. The Grand Anicut was a massive dam, 1000 ft long and 200 ft wide, supplying water to 70,000 acres of fertile agricultural land. The dam is still active, the oldest functioning such structure in the world today.

Dams also provide energy, generated by hydropower. The Industrial Revolution relied on energy from large dams. Half a million watermills supplied electricity to Europe's factories and mines by the middle of the eighteenth century. By 1900, Britain alone had almost as many dams as the rest of the world

(today, there are about 2000 dams in England and Wales, and 800 in Scotland.)

From the nineteenth century onwards, the number of dams exploded across the world. The quest to shackle the waters themselves and bend them to human will became central to grand projects of development, fuelling colonial expansion and the growth of modern cities. Before humans began to interfere with the movement of rivers, by building giant dams, the rivers of the world transported 15 gigaton of sediments and silt to the ocean each year. The expansion of dams from the nineteenth century traps more than 15 per cent of this silt, a staggering 2.3 gigaton of sediment, in reservoirs each year. This silts up dams, reducing the quantity of water they can store and increasing the weight they hold, making them seismically unstable. It also erodes the delta regions along the coast, which rely on the silt brought in by rivers to counteract the forces of coastal erosion. Today, large deltas are sinking across the world at a rate four times as high as that of sea level rise. Thus, deltas which already contain some of the world's densest populations in large cities will flood faster and face more intensive floods than we anticipate, as scientists from the International Geosphere-Biosphere Programme have warned.

Because of dams, some studies estimate that between 74 and 95 per cent of the sediment in the Mahanadi, Cauvery, Sabarmati, Krishna and Narmada rivers does not reach the delta. This has already led to the erosion of 76 sq. km of area in the Krishna and Godavari delta area between 1965 and 2008. Excessive harvest of groundwater can also lead to the subsidence of cities. Indonesia's capital city, Jakarta, is sinking at the average rate of 11 in. a year because of the overharvesting of groundwater. The Indonesian government may need to move its capital to another location: along with its 11 million residents. Not an easy task!

India is the third-largest dam builder in the world, after China and USA. About 1100 of the country's large dams are more than fifty years old, needing desiltation and repair. A series of new dams is planned along the rivers of North-east India. New proposals seek to construct twenty-six mega-dams, even to divert the waters of the mighty Brahmaputra into the Ganga. Hydrologists have warned of the dangers of large-scale hydroengineering on seismic stability. In many parts of the country, there is substantial local opposition to these projects by local communities, who fear that the dams will impact their livelihoods and lead to their displacement.

Of India's mega-dam projects, none has perhaps been mired in conflict as much as the Sardar Sarovar Dam. Constructed on the Narmada River, the dam was intended to supply drinking water to cities in Gujarat. Massive protests against its construction were witnessed, beginning in the 1980s, by Adivasis, farmers, academics, NGOs and activists who came together under the umbrella of the Narmada Bachao

Andolan (Save Narmada Movement). The World Bank withdrew its support for the project after an independent review in 1992, saying, 'We think the Sardar Sarovar Projects as they stand are flawed, that resettlement and rehabilitation of all those displaced by the Projects is not possible under prevailing circumstances, and that the environmental impacts of the Projects have not been properly considered or adequately addressed.' The dam went ahead.

Till date, we have no accurate numbers of those displaced by dams—estimates vary from 21 million in the late 1980s to 33 million displaced by the twenty-first century. No one has kept a complete count. In 1949, it was estimated that 168 villages would be submerged by the Hirakud Dam. By 1957 when the dam was completed, 325 villages containing 26,501 families had been displaced. In the case of the Sardar Sarovar, 6000 families were estimated to be affected in 1979. In reality, activists estimate that the dam impacted and displaced 85,000 families, about five lakh people. Many of the displaced families come from especially poor and vulnerable backgrounds, often already landless and therefore not eligible for compensation. Adivasis constitute only 8 per cent of the population of the country. Yet they constitute 59 per cent of the 11.6 lakh people displaced by twenty large dams in the 1990s.

Where resettlement has been promised, land may be given years after displacement. When the

Tungabhadra Dam was built, families were resettled five years after they lost their homes. Many dam refugees are forced to migrate to cities, working as migrant labour, eking out a marginal living. Oral History Narmada, an archive created by oral historian Nandini Oza, records the histories of the Narmada movement. In her words, 'These voices of resistance and loss will help influence the development planners to plan projects that have a smaller ecological, cultural, social, economic footprint.'

We often hear the word 'cost-benefit' in the context of development projects such as dams. It is taken that the benefits of the projects to the nation and its citizens far outweigh the costs of building the dams. But costs encompass more than the financial expenditure on construction, and benefits should encompass more than an economic accounting of return on financial investment. The incalculable human cost of displacement, which falls on the poorest and the most vulnerable, is missing in this kind of accounting. It is all very well to say that one cannot make an omelette without breaking eggs. But in this case, the one who eats the omelette is not the one who lost their eggs, or farms. Like the pregnant woman whom Sir Mirza assisted, people in cities—especially women—benefit from the easy availability of water from taps. But hidden in the costs of bringing water to cities from distant large dams and reservoirs is the question of social justice. Who pays the price for whom?

Dams also come with hidden ecological, climatic and hydrological costs. They impede the natural flow of rivers, impacting the migration of fish and destroying critical breeding and spawning ground, resulting in the widespread loss of biodiversity. A report by historian Ramya Swayamprakash, for the South Asia Network on Dams, Rivers and People, describes the disappearance of the hilsa fish from the Cauvery with the construction of the Mettur Dam (the Stanley Reservoir). The hilsa have also disappeared from local memory—as Swayamprakash describes, even local fishermen have no idea that the river once contained hilsa, or other indigenous fish like the Nilgiris barb. They are only aware of fish like the catla, artificially introduced to the reservoir. Less than a century after the Mettur Dam was built, an entire native ecosystem has been lost—along with local memory of its existence.

Dams are often believed to be good for the climate because they produce renewable energy via hydropower generation. But dams also have a strong impact on climate change, increasing methane emissions. The sediment and organic matter trapped behind the walls of the dam in the reservoir are insufficiently aerated—they have low levels of oxygen because they lack flowing water. This creates anoxic zones, dominated by anaerobic bacteria (bacteria capable of living in environments with low levels of oxygen), which produce methane instead of carbon dioxide.

Methane is a much stronger greenhouse gas compared to carbon dioxide, trapping much more heat and resulting in a far greater impact on global warming. The National Institute for Space Research in Brazil estimates that large dams are responsible for methane emissions of about 104 million metric tonnes per year, which accounts for about 30 per cent of the global methane emissions from anthropogenic sources. A 2016 study published in the journal *Bioscience* reported that that dams contribute to 0.8 petagrammes (billion tonnes) of carbon dioxide equivalent emissions per year. That is, as much as 1.3 per cent of total global emissions are caused by dams. Even this number, the researchers warn, is likely to be an underestimate. Surprisingly, 10 per cent of dams emit more greenhouse gases than a conventional fossil fuel gas-powered plant of the same capacity! The situation is far worse in tropical countries like India, where warmer temperatures and the greater biomass content of the silt contribute to increased methane emissions. Yet many people, and policymakers, persist in thinking—incorrectly—that hydropower is a low-emission solution to climate change.

In Jawaharlal Nehru's famous speech, given at the inauguration of the Bhakra Nangal Dam in 1963, he said,

This Dam is not meant for our Generation alone but for many generations to come as well, who will

derive benefits from it . . . This Dam has been built
up with the unrelenting toil of man for the benefit of
mankind and therefore is worthy of worship. May
you call it a Temple or a Gurudwara or a Mosque,
it inspires our administration and reverence.

But even the best-built dams do not last for 'many
generations to come'. Several countries have begun
to decommission dams, especially older ones that are
now desilted, non-functional and unsafe. The river
conservation organization, American Rivers, states
that over 1956 dams have been taken out in the USA.
Fifty-seven of these were removed in 2021, allowing
close to 3500 km of river length to return to their
natural patterns of flow, unencumbered by reservoirs
and dam walls. The removal of the Edwards Dam
on the Kennebec River in Maine in 1999 marked a
turning point for the USA, paving the way for many
non-functional dams to be decommissioned. When
the dam was constructed, it destroyed one of the most
important breeding and spawning areas for Atlantic
fish. The removal of the dam led to the return of
cascades of water and rapids, with the restoration of
native ecosystems. Local fish species reappeared after
many years, promoting the recovery of native bird
populations.

The dams removed so far represent less than 2 per
cent of all dams in the USA, and there is still a long
way to go. India, which has some of the oldest dams

in the world, may need to follow the same path. In October 2021, the Government of Kerala submitted a note to the Supreme Court asking for the 126-year-old Mullaperiyar Dam to be decommissioned—but only so that a new dam, more structurally sound, can be built. Yet as organizations like the South Asia Network on Dams, Rivers and People point out, there are other approaches we could consider—including restoring the network of interconnected small tanks and canals that once existed, using approaches of watershed planning.

Cities cannot place infinite demands on large dams and reservoirs without leading to environmental and ecological damage—which will, in turn, impact the survival and quality of life of people in cities. Perhaps we should explore a return to the approach taken by the builders of Dholavira, investing in a series of small dams to harness the rainwater. Alongside, we need to revive the practice of using recharge pits and open wells connected to tanks, lakes and ponds, to increase local water supply, restoring wetlands alongside these water systems. Such an approach would be climate-friendly, good for the environment, and easy to maintain, helping cities increase their resilience to floods and droughts.

A model city to emulate could be Wellington, the capital of New Zealand. When a local dam was no longer needed, the city decommissioned the dam, creating an urban eco-sanctuary called Zealandia. Covering 225 ha, located 2.5 km from the city, the reserve is protected by a fence to keep out invasive predator species. Zealandia is a spectacular learning environment for the city, hosting hundreds of native plant species, above forty native bird species and many more reptiles, frogs and invertebrates, many of which are critically endangered. The site is a living laboratory for local residents to connect with nature, with hundreds of volunteers, including schoolchildren and college students, helping at the site and giving guided tours. The reserve has a 500-year plan for restoration, with artists providing visions of the landscape as it might look five centuries in the future at landscape vantage points. For a dam set up in 1878, and decommissioned in 1997, Zealandia has come a long way in restoring the connection between a city and its environment. Many Indian cities could follow suit—if we could only open our minds to more imaginative possibilities.

The Oldest Dams in the World

Ancient civilizations have relied on dams for millennia. Eight thousand years ago, farmers in Mesopotamia, at the foothills of the Zagros Mountains, built irrigation canals, which may have been dammed using temporary structures of mud and wood to direct the flow of water. Another large dam, the Sadd el-Kafara, built by the Egyptians near Cairo about 5000 years back, was a massive failure, breaking down soon after it was built.

But the earliest concrete evidence of a functioning dam comes from 3000 BCE, in the town of Jawa, in today's Jordan. The Jawa Dam system included a 200-m low dam or weir which diverted water from the river into a canal, and then into ten reservoirs, one held by a dam that was 80 m long and 4 m high. The Egyptians, Assyrians, Babylonians, Persians and Sabeans also built a number of dams in different parts of the world between the seventh and the third century BCE, using the water for irrigation. By the first century BCE, dam construction had expanded to include the Middle East, Mediterranean, China and Central America.

Sri Lanka's Royal Dam Builders

The ancient rulers of Sri Lanka (formerly Ceylon) were inveterate dam builders. By the third century BCE, they had paved the way for dams, by inventing the technology of the valve pit, which was used to regulate the flow of water from reservoirs. In the third century BCE, the great king Mahasena built a 40-km canal from the Mahaweli River towards the northern dry zone of the island country. He created the giant Kantale Tank that covered 4660 acres, held by a dam that was 50 ft high. In the fifth century BCE, King Dhatusena built an even larger dam, with an 87-km canal and a tank that was 18 sq. km in area. By the twelfth century CE, dam building in the island country was out of control, with the legendary king Parakrama Babu commissioning the construction of 4000 dams, much like other kings commissioned gargantuan temples. One of these dams was 15 m tall and almost 14 km in length, so large that it remained the biggest dam in the world until the twentieth century. Why did such a small island need so many large dams? Contemporary scholars believe that these mega-dams were only for show, used to reinforce messages of kingly power and prowess. They were less likely to be used by local villagers, who relied on small dams for most of their water.

ELEVEN

UDAIPUR: CITY OF LAKES

When full, this lake, submerging the bunds of two others, throws out deep bays into the suburbs, a picturesque bridge unites one of these to the city, and the sparkling water on either side is edged with numerous ghats, gay balconies and temples, shaded with dark foliage.

Pichola Lake, surrounded by deep bays and picturesque bridges, was a patch of paradise in the arid, barren landscape of Rajasthan—a tribute to the power and majesty of the kings of Udaipur.

Udaipur was once part of the kingdom of Mewar, in the south of Rajasthan. Embedded in a valley of the Aravalli Hills, the city was founded in the sixteenth century. The Rajput ruler Rana Udai Singh fled the advancing armies of the Mughal emperor Akbar, taking refuge in the mountains where he established his new capital. Early descriptions and paintings of the city show us that the lakes and the hills are key defining features of the landscape.

Rana Udai Singh selected a location to the east of Pichola Lake, to build his new city. He later constructed a large dam on the east, creating a reservoir called Udai Sagar. As Udaipur grew, new water bodies were added on the north side of Pichola Lake—Amar Kund, Rang Sagar, Swaroop Sagar and Kumhariya Talab. The ecological landscape of this natural valley was exploited to create these reservoirs or lakes, and supply water to the city from its two perennial rivers—Ayad and Sisarma. The lakes of Udaipur, surrounded by handsome havelis and ghats, with tall trees and dense vegetation, became the backdrop for a courtly, pleasure-centred environment, nurturing local traditions of music, dance, art and food.

Raj Samand Lake, initiated by Rana Raj Singh I between 1662 and 1676, is one of the oldest-known famine relief works in Rajasthan, built to provide employment and income to help those affected by the local famine of 1662. The lake was later used as a base for the British Imperial Airways to land sea planes in the Second World War. Jai Samand Lake, south-east of Udaipur city, was built between 1685 and 1691 by Rana Jai Singh, by constructing a dam across the perennial Gomti River. The lake provided water for irrigation and fishing. In a *Rajputana Gazetteer* in 1908, Major K.D. Erskine described the lake in poetic language, saying that the 'wooded islands and picturesque fishing hamlets on the northern shore add greatly to the beauty of what is one of the largest artificial sheets of water in the world'. In the early 1900s, the water from the lake was diverted via canals to villages around Udaipur, irrigating close to 50 sq. km of agricultural land.

Within the bounds of Udaipur city are the Pichola and Fateh Sagar Lakes, two iconic waterscapes that comprise the cultural, social and economic core of the city. Tourist footfalls are thickest in this part of Udaipur. In the vicinity of these two lakes are many others—Rang Sagar, Swaroop Sagar and Dudh Talai—as well as several majestic stepwells and open wells. Sitting at one of the picturesque ghats next to the lake, one can admire the waterfront set against the backdrop of the Aravalli Hills, with the grandeur of the sandstone forts and palaces, steeped in stories of local heroes such as the Rajput ruler Maharana Pratap.

Legend has it that Pichola Lake was built by a *banjara* (gypsy) who transported grain. Finding that his bullocks were unable to cross a stream, he built a dam to create a pathway. According to historical accounts, though, Pichola Lake was built in the sixteenth century by Udai Singh. It has two island palaces, Jagmandir and Jagniwas. When the Mughal prince Khurram (later known as Shah Jahan) rebelled against his father Jahangir, he sought asylum in Udaipur. The Jagmandir Palace on the island was built to house the refugee prince and his family. The gardens in Jagmandir were dense with fruit-bearing trees such as the mango, banana and orange, while tall cypress and palms added to the charm, with their branches gracefully drooping over the waters. Prince Khurram was said to have been so impressed with the architecture of the palace that he copied elements of its style when he built the Taj Mahal. During the Indian rebellion of 1857, Jagmandir became a refuge once again, this time for English officers and their families.

James Fergusson, the acclaimed nineteenth-century historian who travelled extensively in India, said that the islands on Pichola Lake surpassed the grandeur and beauty of lakes in Italy. According to him, 'Indeed, I know of nothing that will bear comparison with them anywhere.' But the lakes were not built just as places of beauty. Pichola Lake supplied drinking water and was used for bathing. One of the stepped ghats, Lal Ghat, was built for women to bathe privately, away from public view. The elaborately carved buildings around the lake were used for public entertainment, to feed wild boars, watch fights between boars and tigers, and to enjoy the scenic views. In 1795, the embankment of Pichola Lake was broken after a spate of heavy rainfall. Nearly half the inhabited part of Udaipur city was washed away.

Nearby lies Fateh Sagar Lake. Connected to Pichola Lake, it is said to have originally been a smaller tank called Dewali. It was expanded by Maharaja Jai

Singh in the late seventeenth century. After the lake was breached in a storm in the late nineteenth century, the embankment was strengthened and rebuilt by Maharaja Fateh Singh, after whom it is now named. Both lakes are supplied by the waters of the Kotra River, a tributary of the Berach. Two other lakes— Rang Sagar and Swaroop Sagar—were also connected to this water system, and together with stepwells, they supplied drinking water to Udaipur. In 1931, piped water began to be supplied to the city from Swaroop Sagar.

The lakes of Udaipur are star tourist attractions that support the city's economy. Jagniwas, today known as the Lake Palace, is one of the most expensive luxury heritage hotels of India. Other havelis on the lake, such as Bagore ki Haveli on Pichola Lake's Gangaur Ghat, are favourite tourist spots, as are the islands. Tourists can avail of exotic experiences like camel rides, and more prosaic offerings such as popcorn and roasted peanuts. In recent years, the city has become a location for destination weddings of Indian and foreign celebrities.

The commercialization of this lake city, which once supported its growth, now threatens its survival. When the trees that surrounded Fateh Sagar were cut down after Independence, the large heronries near the lake collapsed in size. In the 1950s, contracts were given out for the killing of marsh crocodiles, whose skin was used to make leather products. Within a couple of years, almost all the crocodiles in the lake disappeared. Indigenous varieties of fish gave way to commercially stocked exotic fish, once the lake was leased out to commercial fishing contractors. The marshy wetlands, which acted as bird nesting and roosting sites, dried up when the water works department pumped large quantities of water from the lakes to supply the city. Mechanized boats used by tourists began to contaminate the water with oil spills. Despite interventions by the Rajasthan High Court and a government ban on further construction at the edge of the lake, not much has changed.

The Dewas project was initiated in 1968–69 to supply drinking water to Udaipur and a few other villages, by bringing water from the Sisarma River into Pichola. Despite getting water from Pichola and Fateh Sagar, Udaipur experienced a major drought during 1988–89. Water was supplied from Jaisamand Lake but proved insufficient. By then, the quality of water supply had also deteriorated. The once-clear waters of the Ahar, which supplied the Fateh Sagar, Pichola and other lakes from the Aravallis, had become filled with industrial and household waste, converting the canal into a *nala* or drain. The Mansi Wakal Scheme was proposed in 1989 to address these issues, by building a dam and reservoir on the Mansi River, supplying treated water to Udaipur. Local villages opposed the project, which would submerge many homes, displacing an estimated 6800 people. Despite the opposition, construction on the dam began in 2001 and was completed by 2005.

But Udaipur's thirst has not been quenched. An exponential growth in industries at the city periphery and the increase in luxury tourism, with a demand for swimming pools, jacuzzies and bottled water, have placed enormous pressure on the city's once pristine water bodies. Drains now flow into the lakes whose waters were once directly consumed. Garbage is dumped into the lakes of Udaipur from the homes, hotels and restaurants that now surround them. The network of lakes, the very reason that attracts tourists to the city, is now threatened by the industry it supports.

The destruction of the Aravalli hill ranges around Udaipur also threatens the city—despite being less recognized. The Aravallis act as a geographical barrier to the Thar Desert, and arrest its spread eastwards into the Udaipur plateau. They act as a catchment area that feeds the waters from the rain into the rivers and lakes of Udaipur. Until the early twentieth century, the Aravallis were 'well clothed with forest trees and jungle affording shelter to tigers, bears and panthers, and the scenery is wild and picturesque'. But the *Report on the Administration of Mewar State for the Years 1940, 1941 and 1942* (1944) states that 'now Udaipur is surrounded by hills bare of vegetation . . . Thousands of tons of silt have been deposited in these lakes. There could be no doubt that in course of time these lakes would be silted up completely'.

More recently, an even more dangerous problem confronts the city—the loss of the Aravallis. In a Supreme Court order dated 23 October 2018, the judges observed that thirty-one hills and hillocks of the Aravallis had been completely destroyed by mining. Soil from the denuded hilltops flows down in the monsoons, silting up the lakes of Udaipur. Industrial and sewage effluents that flow into the lakes have increased their nitrogen and phosphorus levels, stimulating the growth of vegetation, pathogenic microorganisms and harmful microalgae. Without the Aravallis, there is a risk that the monsoon rains will decrease, and the rivers that feed the lakes of Udaipur will run dry. These hills act as a barrier to the Thar Desert. If they are threatened, the desertification of Udaipur becomes a real possibility.

An ancient city which flaunted its beauty in paintings and music, and a world destination for celebrity weddings, the grandeur of Udaipur is now a pale shadow of its past. The question for Udaipur today is: can they do tourism right? Or is it just a matter of time before unsustainable growth turns this landscape, once renowned for its lakes and natural beauty, into a dust bowl surrounded by denuded mountains?

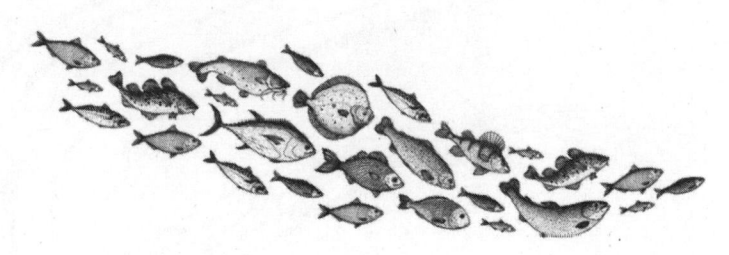

TWELVE

SONGS OF THE RIVER

In the 1940s, a song composed by an unknown young man helped the mega-dams of the Columbia River overcome stiff local resistance. Woody Guthrie—for that was his name—went on to become one of the most famous musicians of USA.

The Bonneville Power Administration (BPA), a hydropower company owned by the US government, was in a bind. Local industrialists and the media opposed the project, and local residents refused to buy power from the dam. In 1941, the director of the BPA interviewed a skinny young songwriter. Pleased with his music, the director commissioned the artiste to compose songs about the river and its mega-dams.

Guthrie, the young man hired by the BPA, composed twenty-six songs on the Columbia River. 'Roll On, Columbia, Roll On', one of his most famous songs, praises the ambitious technological vision that created the Grand Coulee Dam, calling it 'the mightiest thing ever built by a man / To run these great factories and water the land'.

Water quenches our bodily thirst but music and song nourish our souls. Guthrie's paean to the Columbia River became the official folk song of Washington State, helping swing the mood of the public towards large public projects. He was only following in the footsteps of many musicians who composed songs about the water, in different parts of the world.

The boatmen of the rivers of the Upper Yangtze in China plied their oars to the rhythmic beat of the gong. Taking their boats on journeys that sometimes covered over 600 km on the river, they sang an impressive variety of songs. Many songs described the daily hardships of their lives, steeped in manual labour. In others, the boatmen recast themselves as heroes of legend, who personified the best values of brotherhood and community spirit.

The music of the river is the realm of boatmen, bards, philosophers and kings. *Water Music*, one of the most famous western music compositions on a river, was commissioned by King George I in 1917. George Frideric Handel created a set of pieces with spirited, fast-moving music for a royal cruise along the Thames River. Fifty performers crammed into one boat, setting off on a 5.5-km journey. The king and his party of aristocrats listened to the performance from the royal barge, surrounded by boats that filled the river. Historians of music speculate that Handel carefully selected the musical instruments used in *Water Music*. He used wind instruments like the horn, oboe and bassoon, and woodwind instruments like flutes and recorders whose sound carried across the water. His original composition seems to have avoided the use of other popular instruments of the time such as the timpani, or kettledrum, whose deep vibrations would have dissipated on the water, making it difficult for the other boats to hear their sound.

With his long experience of music, Handel must have known which musical instruments to use on the water. But birds, who lack access to musical instruments, have an intuitive knowledge of this. They modulate the frequency of their calls to ensure their voices carry across the water. Active mountain streams produce a lot of low-frequency noise. Birds like the whistling thrush found next to such streams sing at a high pitch, making sure their songs carry above the sound of the moving water. Chaffinches find another way to cope. When they are near noisy waterfalls or falling water, they repeat their song multiple times, ensuring that it is heard. Evolution is truly marvellous.

Some songs flow gently, like the ripples of a slowly moving stream, bringing comfort and healing. Others bring to mind the roar of rushing water in a waterfall. A song of the river may be literal like 'Roll On, Columbia, Roll On', describing a physical aspect of a water body. It can also be allegorical, relating to the human condition.

Oh re, taal mile nadi ke jal mein.
nadi mile saagar mein.
saagar mile kaun se jal me?
Koi jaane naa.

This haunting song from the 1968 Hindi movie *Anokhi Raat* asks a simple, yet deeply philosophical question. The pond flows into the river, the river into the ocean. Where do the waters of the ocean flow? No one knows, says the writer, pushing us to think about the journey of the human soul through birth, life and eventually death. What comes after death?

It is not just the majestic dams and large rivers of the world that inspire songs about the meaning of life and the human condition. Kabir, the fifteenth-century mystic whose *doha*s (couplets) are famous across the country, used the illustration of a small well to mock the narrow vision of humanity, seeking to divide people on the basis of caste and religion. There is just one well, he said, and its waters are one. But each one who comes to the well, to fill their vessel, seeks futilely to establish their differences.

Kabira kuan ek hai, pani bharain anek
Bhaande hee mein bhed hai, pani sub mein ek.

Water music of different kinds can be found in India too—in traditional folk songs, in commercial films, and as a part of spiritual and sacred music.

Songs of the Ganga and the Yamuna are Bollywood favourites.

Tu Ganga ki mauj, main Yamuna ki dhara
ho rahega milan yeh hamara tumhara.

In this well-loved song from *Baiju Bawra* (1952), the singer says to his sweetheart: 'You are the waves of the Ganga, and I am the waters of the Yamuna. We are fated to unite.'

In contrast to this gentle-paced song, is the fast-paced '*Yeh chand sa roshan chehara*'. Who can forget this song, with Shammi Kapoor courting a shy Sharmila Tagore, comparing her face to the beauty of the moon that shines on the glorious waters of the Dal Lake? Another unforgettable Bollywood song is '*Chingari koi bhadke*' from the movie *Amar Prem* (1972). This melancholic song, in which Sharmila Tagore plays a kind-hearted sex worker, and Rajesh Khanna, a lonely businessman, seems to be shot on the Hooghly River. But because the film crew did not get the necessary permissions, it was actually recorded in the famous Natraj Studio of Mumbai. The twinkling lights of the Hooghly, and the silhouette of the famous Howrah Bridge were later transposed on to the film to lend it legitimacy. Perhaps this explains why the actors seem so unconcerned, when the shot seems to depict unruly waves of water lapping at the sides of a shallow, unstable boat.

In Telugu films, the Godavari River takes the place of the Ganga. '*Goddari gattundi, gattu meeda chettundi*' says the superhit song from the award-winning film *Mooga Manasulu* (1964), describing the Godavari, with its bank, and the tree on the riverside. *Godavari* (2006), another Telugu movie, is a short, sweet movie featuring a romance on a river cruise, set against the scenic backdrop of the riverbank and hills.

The land around the Cauvery was fertile ground for music. Many famous composers of Carnatic music lived on the banks of the river. '*Taaraka bindige*', a famous *krithi* (musical composition) that is popular even today, was composed by Purandara Dasa, the fifteenth-century poet-composer-philosopher. The song beautifully evokes a different time of the past, where everyday life centred on the water. The singer pleads with his sister to bring a pot to collect water, water that is like *amruta* (nectar), with the *rasa* (essence) of the Hindu god Rama, at the place where beautiful young women gather. He seeks to use the water to worship God Vittala, becoming one with him. Another Tamil saying aptly puts it: 'To be on the banks of the river Cauvery while drinking in tunes of the Raga Saveri, is to taste eternal bliss.'

The eighteenth-century musician Thyagaraja, a member of the renowned trinity of Carnatic music composers (along with Shyama Shastri and Muthuswami Dikshitar), had an especially close relationship with the Cauvery. One story narrates that Thyagaraja's brother, annoyed with him for refusing to perform in the court of the king of Thanjavur, throws Thyagaraja's precious idol of Rama into the river. A distraught Thyagaraja is said to have composed his well-known krithi 'Nenendu vetukuduraa' [Where shall I search for you, O Lord], while searching for the idol in the river. The Cauvery, so important in the life of Thyagaraja, plays an equally important role in keeping his memory alive. His samadhi lies in the village of Thiruvaiyaru, where he once lived. Every year, musicians and music lovers gather in thousands on the banks of the Cauvery, in this village, to pay homage to the great composer during the Thyagaraja Aaradhane, the commemorative music festival held in his memory.

Most composers of the Bhakti movement were from higher castes. But even in their accounts, we can see glimpses of caste prejudices that shaped society. Sridhara Venkatesa Dikshitar, known as Ayyaval, was an eighteenth-century Brahmin singer who lived on the banks of the Cauvery, in the village of Thiruvisanallur. He was ostracized by others in his community for giving holy food, cooked for an annual shraddha ceremony (performed in honour of a dead ancestor), to hungry Dalits. The angry priests cursed Ayyaval, demanding that he purify himself by immersion in the waters of the holy Ganga. Undeterred by the criticism, he prayed to Shiva to intervene, reciting the hymn, 'Gangashtakam slokam'. Legend has it that Ganga herself appeared before him, flowing out of the well in his home, and flooding the entire village. This incident is remembered each year on Kartigai Amavasya, an auspicious day in the Hindu calendar, when the Ganga is believed to have entered the waters of Ayyaval's well. Tens of thousands of devotees visit the well to take a dip in its holy waters.

Yet, caste continues to influence the recognition of music. While the compositions of the renowned trinity, and the saints of the Bhakti movement are taught widely, the music performed by subaltern groups, especially by Dalits around water bodies, is omitted from formalized music education. Like the boatmen of Shanghai, the women who live around the lakes of Bengaluru sing songs that document local practices that are fast fading from memory, including songs of festivals and worship, of rape and abuse of women and of human sacrifices at the lake. The songs express gratitude to local goddesses such as Gangamma and Duggalamma, seeking the well-being of the village and thanking the goddesses for keeping the village supplied with plentiful water and protecting them from floods. The songs also illuminate the connections between lakes, picturesquely describing how water dances and flows from the lake upstream to the ones downstream, connecting them in the form of a chain.

Bhupen Hazarika, the renowned Assamese musician, singer and poet who received the Bharat Ratna, one of India's highest civilian honours, is also known as the 'Bard of the Brahmaputra'. He is said to have been deeply influenced by the American football player and singer Paul Robeson, whose songs were anthems for black liberty. Hazarika's songs such as '*Bistirno parore*' used the Brahmaputra to speak about issues of inequality and injustice in society, bringing them to the attention of a wider audience just as Robeson's famous song, 'Ol' Man River' did in the USA.

The Narmada Bachao Andolan, one of India's best-known social movements to protest the construction of dams on the river Narmada, used songs in the movement too. These songs were composed and sung by local Adivasi farmers protesting the construction of the dam, along with activists. They played a crucial role in mobilizing people towards non-violent, peaceful protests.

More recently, songs composed on the water have been used to foreground the impacts of climate change. Communities in North-east India living along the banks of the Brahmaputra have recorded their suffering through folk songs that speak about displacement and the loss of lives and livelihoods, warning people of the perils of ignoring climate change.

In the Island Republic of Vanuatu, a Y-shaped archipelago of islands in the Pacific Ocean, women follow a tradition of water music called etëtung, handed down from mother to daughter for many generations. Sound and rhythm are created by using hands on the water to create different sounds. By stroking the surface, slapping the water hard, poking the water with two fingers, and plunging two fists into the water, women mimic the sounds that surround them—rain falling on stones, dolphins slapping their fins in water, people herding fish into traps, and so on. As an island archipelago, Vanuatu is one of the most vulnerable to the impacts of climate change. Because all it requires is waist-high water, etëtung is becoming a medium to carry the message of climate justice from the islands of Vanuatu across the world.

In cities, water largely serves a utilitarian purpose today. Conveyed to our homes in pipes and taps, the 'water music' of lakes and rivers has lost its importance, with many of the songs once sung even in cities now fading from our memory. Yet all is not lost. Many older women are repositories of this knowledge and heritage, giving us hope that we can keep subaltern traditions alive, even if in a new form. Meanwhile, just like the women of Vanuatu, who create music with just their hands and water, we can turn to instruments like the jaltarang which requires only a set of steel or ceramic glasses with water and a stick, to create beautiful music—and reconnect our minds and souls to the songs of the river.

Songs of the Sea

Blue whales, the largest animals on the planet, communicate with others of their species through low frequency sounds that carry over 800 km underwater. These 'songs', typically at a frequency of 14–50 Hz, are difficult to hear with the human ear. Recent research shows that the frequency of blue whale songs has decreased over the last few decades. Scientists studying these changes at the Scripps Institution of Oceanography in San Diego have an explanation for this. In the 1960s when the earliest recordings of blue whale songs were made, commercial whaling was at its peak, and the densities of blue whale populations were far lower and more dispersed than they are today. Male blue whales, who produced most of these songs, needed to search for longer distances to find potential mates, hence they were forced to call at higher frequencies. Today, with the ban on commercial whaling, blue whale populations have rebounded. Males can find a mate much closer at hand and have reduced their frequencies and therefore the distances over which their calls can travel, to a level that is more comfortable for them. In contrast to blue whales, toothed whales receive higher pitched whistles and clicks through their jaw, transmitting it further to their ear. The bottlenose dolphin rules the high registers, transmitting and picking up sounds at 1,60,000 Hz, which we are unable to hear. They use their songs to echolocate, track and assess the shape of distant underwater objects—a skill so useful that military scientists have put dolphins to work to find submarines, and locate buried mines and explosives.

THIRTEEN

INTERLINKING RIVERS

Along the coast, when the monsoon rains are too heavy, sea levels rise. Saline water enters the canals and freshwater river channels, making the once-fertile agricultural land next to these waterways salty and unusable. At other times, the choked and swollen rivers burst out of their banks, leading to avulsion—i.e., changing course to a new location. In doing so, they can submerge entire villages, like the 2008 floods in the Kosi River that displaced nearly three million people.

As schoolchildren, we were taught that 'water finds its own level'. In response to the problems described above, river interlinking seems like an engineer's dream project, a permanent way out of India's recurrent challenges of flood and drought. The underlying idea seems simple—some parts of India have rivers with a water surplus while other regions are parched, with seasonal rivers that run dry in the summer. By connecting all of India's rivers into one massive network, river linking enthusiasts believe that we can expand irrigation across India, minimizing the threat of floods and making drought a problem of the past. They argue that river interlinking will enable the expansion of hydropower plants on many more rivers, supply water to cities and industries, and transform riverways into transport channels—enabling boats and ships laden with cargo to reach far-flung corners of the country.

The National River Linking Project, an ambitious country-wide programme, plans to create sixteen links between peninsular rivers, and fourteen links between rivers that flow down from the Himalayas. When completed, the project will join the Godavari and Krishna rivers, and the Ganga, Damodar and Subarnarekha, along with many others. Across India, the project aims to build thirty canals extending over a length of 15,000 km, along with 3000 dams and reservoirs to form a massive network, the South Asian Water Grid.

Such a grand plan must stem from a modern vision, you may think. But the first seeds of this vision were laid in 1858. Colonel Arthur Cotton, a hydraulic engineer who had built a formidable reputation based on successful irrigation projects on the Cauvery, Krishna and Godavari rivers, was asked to find ways to prevent the Mahanadi River in Odisha from flooding—something it did far too frequently. Cotton did not confine his suggestions to the Mahanadi. He devised a plan for a mammoth project, one that would connect areas as far flung as Karachi and Chennai, Pune and Kolkata. His project would cost Rs 13 million. However, because these new networks could be used to transport ships and boats, and to expand irrigation, it would generate a 30 per cent profit.

However, Cotton's grand vision remained on paper. He had to compete with a larger, even more ambitious infrastructure project—the Indian railways. Cotton promised that his plan would provide better transportation at a fraction of the cost, but the Raj wanted its railways. Massive investments of capital were poured into the expansion of the Indian railroads, and Cotton left India a disappointed man. His Mahanadi plan collapsed, as did the private irrigation schemes which he devised. But Cotton left something behind that was of far more fundamental importance, even though he did not realize it at the time. Although his ideas for the river interlinking project did not materialize in the nineteenth century, the idea caught the imagination of many. It was a heady vision of an interconnected India through its waterways, one that independent India later became keen to take up.

In the 1960s, the Union Minister for Power and Irrigation, K.L. Rao, himself an irrigation engineer, revived the idea of interlinking rivers. He proposed to build a canal that would connect the Ganga in the north to the Cauvery in the south over a distance of 2500 km.

Though attractive, the project was too expensive, and never really took off. In 1978, an aircraft pilot, Captain Dinshaw J. Dastur, came up with a bigger and better plan. Instead of one canal, why not construct two? His plan imagined a canal that extended over 4200 km in the Himalayas, and another that ran for 9300 km in the south. The two channels would be connected at Delhi and Patna, forming a necklace of waterways.

In a booklet he wrote in 1978, Captain Dastur termed his idea 'The Garland Canal Project: Answer to India's Flood, Food and Unemployment Problems'. All in one fell swoop! He promised it would provide benefits of cultivation because of irrigation, hydroelectric power to fuel the needs of a growing country, solve India's transport challenges and generate employment for the poor. This scheme, like its predecessor, was also too costly to implement. Yet the National Perspective Plan of 1980, prepared jointly by the Ministry of Irrigation (Ministry of Jal Shakti) and the Central Water Commission, also proposed the idea of inter-basin transfer of water from 'water-surplus' to 'water-deficit' areas. A new agency, the National Water Development Agency (NWDA), was set up in 1982 to carry out scientific studies of the peninsular rivers.

The Government of India established a commission with a very long name—the National Commission for Integrated Water Resources Development. In its 1999 report, the Commission said that the requirement of water for irrigation in India was undoubtedly high, but these mega-schemes were expensive and impractical. At the same time, between 1982 and 2003, the NWDA carried out a series of studies on how rivers could be interconnected in different parts of the country. First, the excess water would be stored in dams—either old ones, or dams newly built for this purpose—and then transferred through canal systems to other rivers, sometimes through a new set of dams and reservoirs. This system of interconnection was called a 'link'. They explored the possibility of many such links across the rivers of the Himalayas, and of peninsular India.

Several elected governments raised the issue of river interlinking but one speech was especially influential in getting the Supreme Court of India involved. In his 2002 Independence Day address to the nation, the then President, A.P.J. Abdul Kalam, raised the issue of river interlinking, saying, 'It is paradoxical to see floods in one part of country while some other parts face drought . . . The need of the hour is to have a water mission . . . One major part of the water mission would be networking of our rivers.'

Ranjit Kumar, a Supreme Court lawyer who was also amicus curiae in other cases on river pollution, filed an intervention application in one of these earlier cases. He requested the court to examine the river interlinking scheme that the President of India had spoken of. The Supreme Court took keen interest, converting his application into an independent writ petition. In its eventual suggestion, the court supported the interlinking of rivers. Notably, the court suggested the constitution of a High-Powered Committee, to see that these projects were completed on time. The NWDA had proposed a timeline by which the peninsular rivers would be interlinked by 2035, and the Himalayan rivers by 2043. But the court felt that this was too long a time frame, saying the entire project could be completed in a decade. In 2012, giving its final decision, the court directed the government to create a special committee to develop guidelines and oversee the project, once again stressing that it needed to be completed within a definite time span.

In the years that passed, several government reports have been published, examining the specifics of each link project. Yet most of these remained classified, not released for public or expert scrutiny. As a result, there is a lot that remains unknown to the wider public. Meanwhile, heated debates have worked their way into newspapers, news channels and social media. Civil society groups working on water, as well as engineers and scientists, express strong reservations about the project, some also terming the court order an instance of judicial overreach. The plan has also met with resistance from some state governments.

Why such resistance for a project whose underlying logic seems quite simple and obvious?

The first problem is the question of governance. To whom does the river belong? Unlike a forest or a lake, a river is not a stationary object. It flows down a mountain, making its way into the gentler valleys and flat plains, eventually emerging at the delta before it joins the sea. On its long journey to the sea, the river passes through many countries, states and districts. Take the case of the mighty Brahmaputra, the highest point in the Himalayan river interlinking project. The Indian plan for river interlinking begins at a point in Arunachal Pradesh. From here, the waters of the Brahmaputra will be diverted through a series of other rivers—Manas, Sankosh and Teesta—making their way into Nepal, and then back to India, swelling the waters of the Ganga. From the Ganga, this excess water will then be diverted into another series of canals, eventually ending up in the south of India.

This news led to alarm in downstream Bangladesh. The Ganga-Brahmaputra delta discharges into the sea in the Bay of Bengal. Two-thirds of this delta lies in Bangladesh. If the waters of the Brahmaputra were diverted to the south of India, what would become of Bangladesh?

While such discussions were ongoing, China announced its plans to divert the waters of the Yarlung Tsangpo in Tibet to irrigate the dry north-western region. The Yarlung Tsangpo is the main tributary that feeds the Brahmaputra. If this river were to be diverted by China, then the part of the Brahmaputra in Arunachal Pradesh would be considerably drier, leaving little surplus for the rest of India, let alone Bangladesh. The upstream country always has the advantage.

So does the upstream state.

While water-receiving states like Tamil Nadu, Rajasthan and Gujarat have been historically supportive of the idea, some water-providing states, like Assam, Sikkim and Kerala have opposed it at various times, arguing that they should have the right over the water that passes through their territory. The question of legal jurisdiction is complex. In the Indian constitution, water resources are the purview of the state but the Centre has jurisdiction over interstate rivers.

Specific link projects have an even more complex history. Maharashtra has been unwilling to proceed with the Par-Tapi-Narmada link, which aims to take water from the rivers of the Western Ghats, in north Maharashtra, to the drier areas of Gujarat, in Saurashtra and Kutch—a project which has also been opposed by Adivasi groups in Gujarat who fear displacement due to dam construction.

Punjab and Haryana have also been engaged in a long dispute for several decades over the construction of the Yamuna-Sutlej link. This project aims to take water from the Sutlej in Punjab to feed the water-starved Yamuna canal in Haryana. The dispute made its way to the Supreme Court, with Punjab arguing that the state, which is suffering from its own water crisis, cannot spare any water for other states.

There are also fundamental scientific concerns of river hydrology and ecology. At its core, the river interlinking project is based on the idea of making better use of 'wasted' water. But, if you take river hydrological flows into account, water that flows into the sea is not 'wasted'. The silt that is carried by these waters is used to build up the deltas, which would otherwise sink due to constant erosion by the sea. The fertile silt is used by farmers, who have a more holistic view of the river than many engineers, appreciating the river as a living entity. As one farmer whom we interviewed, living on the banks of the Yamuna in Delhi, said, 'Through floods, we regain our lost relationship with Maa Yamuna, and Maa Yamuna reclaims her encroached lands and shows her wrath.'

Damming and channelling rivers to remove 'excess water' will create large channels of stagnant water, an approach which has been known in many past projects to increase salinity. When water flows through channels that are thousands of kilometres long, under the heat of the sun's rays, there will be huge losses from evaporation. Stagnant water in canals also promotes mosquito breeding, increasing the spread of vector-borne diseases like malaria and dengue.

River interlinking presents other ecological challenges. When rivers are interconnected, the plant and animal species from one river, adapted to that environment, will flow into another kind of river, altering its species composition and ecology. Past experiences tell us that some local species will go extinct, while other species may invade the new habitats they encounter, becoming invasive and changing the riverine ecology, as well as destroying local biodiversity and fish species.

India has implemented several smaller inter-basin water transfers, including the Periyar-Vaigai project of 1985 and the Beas-Sutlej link, built in 1983. But this grand river interlinking project, if implemented, will be one of the world's largest inter-basin transfer projects. It will regulate five times the water flow managed by all the inter-basin water transfer projects in the USA.

For now, even the first of the thirty planned links is not complete. The Ken-Betwa river link aims to transfer water from the Ken River to the Betwa, through the construction of the Daudhan Dam, and a canal extending over 321 km. It will cover two states: Madhya Pradesh and Uttar Pradesh. When complete, the project will submerge thousands of hectares of dense forest, including parts of the Panna National Park. This project was approved by the Union Cabinet in December 2021, at a cost of close to Rs 45,000 crore. It is expected to take at least eight years to complete, ending in 2029.

Meanwhile in China, the world's largest and most expensive water transfer ongoing project is underway. The South-North Water Transfer Project aims to fulfil a vision of Mao Zedong, China's iconic former President, to transfer water from the water-rich south to the parched north of the country. Hundreds of thousands of people have been resettled for the project, which has faced challenges that include water pollution, increased fish mortality, and major cost over-runs. Interlinking projects have been proposed in many other parts of the world, such as the Acheloos River diversion in Greece, the inter-basin water transfer schemes in Canada and the Korean Grand Canal Project of South Korea. Some have been completed, while many others have been opposed by people's groups and environmental activists.

Proponents of river-interlinking often say that there is no viable alternative. Yet alternate options do exist. The money to be spent on massive interlinking projects could be used to revive and restore water bodies across the country. Small-scale dams, barrages and channels can be created. Compared to large dams and canals that extend for thousands of kilometres, these will be easier to keep silt-free, protect from pollution and prevent mosquito breeding. They would also help restore local ecosystems and foster closer community engagement.

A closer involvement of experts of many kinds—elected leaders, bureaucrats and planners, economists and governance specialists, irrigation and fisheries experts, as well as ecologists, hydrologists, geologists and public health officers, fishers, grazers and other local community experts—is needed to better understand this complex issue. Diving into project execution without getting the complete picture can be a recipe for disaster.

FOURTEEN

CHENNAI: CITY OF FLOODS AND DROUGHT

Apop quiz question for you: how did Leonardo DiCaprio describe Chennai?

Answer: As a city that needed saving—by the rain.

In June 2019, the Oscar-winning actor posted a picture on Instagram. The photograph showed a group of women, standing around a dry well, trying to access water. 'Only rain can save Chennai from this situation,' said DiCaprio. Perhaps he was not aware of the Chennai floods of 2 December 2015.

From December 2015 to June 2019: in three years and six months, Chennai transformed from a flood-ridden disaster zone to a city that was bone dry. By the end of 2015, water had flooded the roads of Chennai. Boats were taken on to the roads, to rescue the stranded. By mid-2019, Chennai was forced to use a goods train to import drinking water from other towns.

It seems absurd that a coastal city, which receives plenty of rain, should go dry in the summer. And equally bizarre that a city which was once filled with lakes and wetlands, that absorbed the excess water of the monsoon, would flood so completely after a few days of rain. Sadly, Chennai is not unique. The story of this once-marshy city mimics the state of many of our other coastal cities, locked in a seemingly endless cycle of drought and floods. Unless our cities learn how to break this cycle, they cannot lay claim to a secure future.

The south-west monsoons between June and September provide India with 75 per cent of its rainfall. Many parts of India, including Chennai, also receive rainfall from the retreating south-west monsoon winds, beginning in mid-October. This monsoon is sometimes characterized by cyclones, depressions and cloudbursts, with unusually heavy rainfall.

Flooding rivers bring a range of emotions. *Godaari vacchindi*! The Godavari has come! This is the traditional call of the people on the banks of the river in Andhra Pradesh—a call of excitement, a prayer to the river goddess, endearingly forgiving of the damage she causes. The Damodar and Kosi rivers are known as the Sorrow of Bengal and Bihar, because of the destruction they cause when they flood. These floods seem distant to many of us, who interact with them only via the television screen, viewing their images and the associated humanitarian crisis at a safe distance.

For the residents of Chennai, though, floods have become a common phenomenon.

The destruction caused by the retreating south-west monsoon of 2015 was not only due to unseasonal heavy rainfall, it was also due to poor planning and mismanagement. The first episode of heavy rainfall took place between 8 and 11 November, inundating many homes. In a few days, the waters receded. But on 1 December, there was a day of torrential rain, bringing 490 mm of rainfall in a twenty-hour period to a city whose average annual rainfall is only about 1400 mm. One-third of Chennai's typical annual rainfall came down in a single day!

The rains in November had filled the Chembarambakkam reservoir. One of Chennai's largest water sources, the reservoir can store 3645 million cubic feet at full capacity. The dam was full. The gates of the dam should have been opened to release water during a break in the rains, when it would have been safe to do so, but this was not done. As the rains resumed, and the water levels rose higher, the government was worried that the reservoir would burst. They opened the gates without warning. Water from the reservoir flowed into the Adyar River, transforming it into a raging torrent, which overflowed into neighbouring channels, canals and marshes that were already full of water.

No city could tolerate such an excess, particularly not Chennai, which had already lost much of its natural cover. In the 1980s, 80 per cent of the city's landscape consisted of wetlands, as a study by Care Earth Trust, an NGO working on lake restoration in Chennai, shows. By 2010, it was less than 15 per cent. Another study by The Nature Conservancy estimates that more than 85 per cent of Chennai's lakes are degraded, affecting the city's capacity to withstand storms, floods and heavy rainfall.

Of course, the city sank. People watched, worried, as the water steadily rose before their eyes, rushing into homes even before they could move their belongings or leave for a safer place. Desperate appeals for help were made and communities got together to launch large-scale rescue efforts across the city. Given the lack of sufficient government response, residents began to use social media to collate information. Tweeted half a million times in less than twenty-four hours, hashtags such as #ChennaiFlood, #ChennaiRains and #ChennaiRainsHelp brought global attention to the floods, helping rescuers identify the most threatened areas, and arrange flood relief and evacuation in boats, army trucks and helicopters. Sadly, many people lost their lives waiting for rescue.

We may think that natural disasters like floods affect us all, equally. But that is never true.

As the incessant rain tore the city apart, the worst damage was wreaked in the poorer parts of the city once again demonstrating how caste continues to play a role in urban India, even in times of disaster relief. The slums on the banks of the raging Adyar were washed away—entire homes disappeared underwater. Meanwhile, as the rains abated and the waters receded, the city had to be cleaned. The sanitation workers whose own homes were destroyed were forced to work, extricating dead bodies, removing bloated carcasses of animals, and cleaning the filth that had accumulated. These workers were temporary employees who could not afford the luxury of saying 'no'—if they did not work, they did not eat. They were given no protective clothing, not even gloves or boots, even while performing the filthiest of cleaning tasks. Their own homes and belongings were destroyed, but despite their heroic efforts to clean the city, they faced discrimination even while accessing relief. It was no surprise that these workers were mostly Dalits, who survived on daily wages. The flood swept away many things, including our empathy for the plight of others.

At least eighty-nine people died, 23.25 lakh homes were destroyed and public infrastructure was left in a shambles. As the city limped back to normalcy, the post-mortem began. So did the blame game. The weather was easy to blame, of course—there *had* been unusually heavy rains. The El Niño effect, which causes the warming of the ocean's surface, played a major role. But discussions in the news laid the lion's share of the blame on the lack of urban planning, especially the destruction of water bodies and wetlands.

By 2019, the same city was confronted with a new crisis—water scarcity.

Chennai is not new to water scarcity. As far back as the seventh century, during the reign of the Pallavas, the region was considered drought prone. When Alexander Hamilton, a Scottish captain and merchant, arrived at Fort St George in 1718, he was surprised to see there was 'no drinkable water within a mile'. Hamilton wondered why such a location was being chosen to build the new city of Madras (as Chennai was then called).

More recently, beginning with rapid urbanization in the 1980s, Chennai faced a series of droughts. In 2003–04, a major water crisis stimulated demands from environmentalists for water conservation and led to a government mandate for rainwater harvesting across the city. Well implemented in the early years, this welcome initiative relieved water stress for some years. The plan however functioned with a very limited understanding of what rainwater harvesting meant. Considering it sufficient to collect rainwater, city planners failed to understand the importance of recharging the city's groundwater aquifers. For this to happen, the lakes and wetlands were needed. But, lacking protection, the city's water bodies continued to be destroyed.

In 2019, newspapers were filled with reports that Chennai was approaching Day Zero, the day the city would completely run out of water. 'Day Zero' is a term that came to the attention of the public after it was used by the South African city of Cape Town during the water crisis of 2018. The city planned to ration water as it approached Day Zero, providing drinking water only at certain collection points in the city and withholding it from household taps. Although the crisis was thankfully short-lived, it popularized the catchy term, making people all over the world pay attention to the extreme water shortages faced by many global cities.

By May 2019, the water situation in Chennai became worrying. After several days of extremely high temperatures, exacerbated by a delayed monsoon, all the reservoirs that supplied water to Chennai—Poondi, Chembarambakkam, Puzhal and Sholavaram—were dry. The wetlands in Chennai, the Buckingham Canal and the rivers that flowed through the city had also dried up. The poor who lived in informal settlements were used to standing in long queues to fill water from tankers each summer. This time, water scarcity began to hit the middle class as well as affluent residents, who had to deal with mandatory water rationing. Hotels and businesses were especially affected, many having to send guests and employees away, even temporarily close. As an emergency measure, the city received trains, each with fifty wagons, bringing 25 lakh litres of water to Chennai from Vellore. The arrival of the rains in a few days eased the situation, but it is only a matter of time before Chennai faces a repeat of Day Zero.

What is the solution? To understand how to make Chennai resilient to floods and droughts, we have to first examine the historical processes of urban growth that led to the widespread neglect of the city's waterscape that we see today.

Well before Chennai, or Madras, became a city, the larger landscape around the region was known to be dry and drought prone. The region was governed by several dynasties in succession, including the Pallavas, Cholas, Pandyas and the kings of Vijayanagara. The area that is Chennai today forms part of the ancient Tamil landscape of Thondaimandalam, a region that extended between the Penna and Ponnaiyar rivers, covering areas as far flung as Kanjeevaram, Vellore and Tiruvallur. Unlike the Cauvery, these rivers were largely seasonal, drying up in the summer. The villages in the area around Chennai needed a reliable source of water during the dry months. The Pallavas built a number of *eri*s (tanks) to collect and store rainwater. The Cholas continued this tradition, maintaining and interconnecting tanks, and building canals for irrigation.

Early descriptions of the landscape where the city of Chennai was founded seem particularly unprepossessing. The East India Company was eagerly looking for a trading post on the east coast, having failed to set up one in Machilipatnam. Despite the sandy, barren appearance of the land, they rented a location on the coast from the Raja of Chandragiri, a subsidiary of the Vijayanagara empire, in 1639 to set up a trading warehouse. Fort St George, established at this site, soon grew into a trading hub. The Madras City Municipal Corporation, established on 29 September 1688, was the first municipal body in India. This paved the way for the growth of the city that held strategic importance for the British Empire.

The British residents of Fort St George depended on well water for their needs. A set of seven wells that once supplied the fort now lie neglected at the edge of the fort, overgrown with weeds and rank with disuse. By the 1690s, the inhabitants of the fort were so many that these wells were no longer sufficient for their use. The British built a channel that connected the large tank of Chembarambakkam to Fort St George. Subsequently, other large wells were dug 3.2 km north-west of Fort St George to supply the growing urban centre with water. In 1773, the city pioneered advancements in urban water supply technology by laying pipelines to bring water into the fort, the first of its kind in India at that scale. The Europeans also received pure water from St Thomas Mount.

In contrast, the native population, living in the area with the derogatory name of 'Black Town' around the fort, had to rely on shallow wells and were often forced to drink contaminated water. By the 1800s, Black Town began to face water scarcity, so additional wells were dug for the use of the 'natives'. As the town grew, wells became insufficient here, too. Water was diverted from the Kosasthalaiyar River in the 1860s into the Puzhal Tank, which the British called the Red Hills Reservoir. From here, the British planned to pump water to Spur Tank, then distribute it across the city, but this project proved to be too expensive. Instead an alternate tank in Kilpauk was used which, at a higher elevation, enabled water to flow towards the city by gravity.

The Madras Municipal Water Works was formed in 1882, in Kilpauk. When the city grew further, the municipal supply from Puzhal was extended to the suburbs. But Chennai did not forget its wells and tanks. A number of small tanks, open wells and temple tanks continued to spread across the landscape, supplying water to local residents. Rao Sahib C.S. Srinivasachari in his book *History of the City of Madras* (1939) writes, 'Madras is honeycombed with hundreds of small tanks.'

In 1978, the management of water supply was transferred to a different organization, known today as the Chennai Metropolitan Water Supply and Sewerage Board. Four reservoirs—Poondi, Puzhal, Chembarambakkam and Cholavaram—now supply water to Chennai. All are located beyond the municipal boundaries to the west of the city, with Poondi at 60 km being the farthest. These tanks are no longer sufficient to supply the large and ever-growing population of the city. Water is also brought in from the Krishna River in Andhra Pradesh, through the Telugu Ganga Project over a distance of 408 km. Another source is from the Cauvery River via Veeranam Lake, over a distance of almost 570 km. But many of the older tanks, such as Spur Tank, Vyasarpadi Tank, Long Tank and Nungambakkam Tank, no longer exist.

What about the three rivers of Chennai—Adyar, Cooum and Kosasthalaiyar? And the Buckingham Canal and marshes of the city including the Pallikaranai marshes? In the early days of Chennai's growth, when the city's drainage was being planned, the network of rivers and the Buckingham Canal were expected to receive and dispose of the excess rainwater, via sewers and drains that connected homes, offices and industries to these water bodies. Perhaps the most important of these for Chennai was the Cooum River, which later joins the Kosasthalaiyar, eventually reaching the sea at the Bay of Bengal. By the early twentieth century, the once-functioning Cooum had become a drain. Fanny Emily Penny, a local resident, describes the river in her book *Fort St. George* (1900):

> The river, confined to narrower limits in the present day . . . is scoffingly dubbed 'The Silvery Cooum.' . . . it has a trick of assuming in the tropical sunset a fascinating beauty and fairness . . . and the eye of the artist is equally delighted as his nostril is offended . . . When the sky is purple with the gathering clouds of the monsoon, the Cooum ruffles its waters . . . the black wet ooze glistens

with delicate shades of pearl. But the Cooum is not remembered for its false and transient beauty; it is indelibly stamped on the memory of the Anglo-Indian by its odours.

More than a century after Penny wrote about the Cooum, the stink of the river has acquired legendary proportions in the local imagination—you can smell the river before you see it.

Like the Cooum, the Adyar River has also become a drain, filled with sewage that enters at multiple points on its journey to the sea, making it compete with the Cooum in terms of its fetid odour. S. Muthiah, the renowned historian and memory keeper of Chennai, once wrote in *A Madras Miscellany*: 'What tragedy we are bringing upon one of the city's most precious environmental assets and a major historical landmark!'

The Adyar River, a historical landmark? It may be hard to believe but historical battles were waged on the banks of the river, which led to the genesis of the Indian Army of today. On 24 September 1746, the troops of the French East India Company fought the army of the Nawab of Arcot, in a pitched fight later known as the Battle of the Adyar River. This was a key strategic battle, part of the first Carnatic War, at a time when the French and British East India companies fought each other, seeking to gain possession of the east coast of south India.

Although small in numbers, the French won the battle, thanks to their superior artillery units, armed with flintlock muskets. The large cavalry troops supervised by Mahfuz Khan, the Nawab's son, stood no chance against this military firepower. The British took note of the arms training provided by the French and set up the Madras Regiment in Cuddalore, a coastal town nearby. This sowed the seeds of the Indian Army of today.

Few remember this history. There have been some attempts to restore the Adyar River and re-establish its connection to the sea. The Adyar Poonga restoration is a remarkable initiative, which has rejuvenated a small stretch of the river, also establishing an eco-park on the riverbanks. Schoolchildren, local residents and tourists visit the park, where they can view signage created by local artisans, carved on locally available Cudappah stone, to learn about the rich biodiversity that characterizes this region. But larger issues have yet to be addressed, including treating the volumes of sewage that pour into the river daily, and dredging and desilting the estuary where it drains into the sea.

Running perpendicular to the three rivers that cut through the city, the Buckingham Canal is another important water body that was built in bits and pieces over many years. In different stretches, and at different times, the canal has been given many names: Elambore River, Cochrane's Canal, East Coast Canal, South Coast Canal and North River. The section that cuts through Chennai is a small section of a much longer canal that extends over a total distance of 796 km, beginning in the city of Kakinada (Andhra Pradesh) and ending in Viluppuram district of Tamil Nadu.

The very first section—named Cochrane's Canal after Basil Cochrane, a wealthy Scottish businessman who financed its construction—was built in 1806. It connected Chennai to Ennore and was used for navigation. Later, Cochrane's Canal was extended towards Pulicat Lake. Eventually, in 1878, the canal was renamed Buckingham Canal, after the Governor of Madras, who ordered an 8-km stretch of the canal to be built between 1876 and 1878. This portion, connecting the mouth of the Cooum to the Adyar, was built during a time of extended drought, as part of a famine relief project. During the nineteenth and early twentieth centuries, the canal was an important waterway, used to transport passengers and local goods, including fish and firewood, on boats. With the expansion of the railways, the canal lost its importance. Sadly, all that is left of the once-grand canal today is its grandiose name. Its waters are heavily polluted by industrial waste and sewage. Many parts of the canal have been nibbled upon, encroached and taken away by urban infrastructure, especially the Chennai Mass Rapid Transit System, which has drastically reduced the canal's width in some places.

Efforts are on to revive the Buckingham Canal, not for transport but for its capacity to maintain water levels by connecting the waterways of Chennai during times of extreme rainfall—an aspect that the Buckingham Canal fulfilled through its connection to the Cooum even in British times.

Apart from the rivers and canal, the marshes of Chennai have also been impacted by ill-considered urban growth. The Pallikaranai marsh, a large floodplain at the southern end of Chennai, once covered over 50 sq. km. The marsh was a hotspot of biodiversity, used for fishing and to cultivate rice. Hemmed in by the coast on the east, and Andhra Pradesh to the north, the city expanded southwards, eating into the marsh. Designated a 'wasteland', it was easier to convert this area into urban land use. The wetland became a dumping site for Chennai's garbage.

By 2002, the marsh had shrunk to a fraction of its original size, covering just about 593 ha. A flood that year acted as a wake-up call to local citizen groups and Resident Welfare Associations, who formed the Save Pallikaranai Marsh Forum. They used stories in the media and studies conducted by local students on the biodiversity of the marsh, stimulating public interest in the revival of the marsh. In 2007, following sustained efforts, they succeeded in getting 317 ha of the marsh protected and declared a Reserve Forest. This paved the way for the restoration of sections of the marsh by local organizations such as Care Earth.

The marsh supports at least 133 species of birds, with as many as 40,000 birds visiting during the migratory season. Yet garbage continues to be dumped in landfills. Almost 36 lakh cubic metres of garbage are already present in Pallikaranai. An additional 2000 tonnes of mixed, unsegregated waste makes its way into the landfill each day, adding carcinogenic chemicals, toxic heavy metals and plastics into the leachate from the dump that finds its way into the marsh, poisoning the groundwater, the fish and the surrounding wildlife. Very recently, in April 2022, the marsh caught fire because of the methane that was generated in the waste dump.

In 2021, the Madras High Court asked the forest department to ban all non-forest activities, relocate the garbage dumps and reclaim land that had been granted to various government institutions and industries. Garbage, however, continues to be dumped in the marsh, as the fire of April 2022 shows. In April 2022, the Government of Tamil Nadu also proposed to get Rs 200 crore from the Green Climate Fund, a global initiative of the United Nations Framework Convention on Climate Change, to restore the marsh, making Chennai more resilient to climate change.

Do such mega projects work? Even if we are optimistic, past evidence shows that we need to be very careful about the potential impacts on the poor. Often, only cosmetic clean-up of the environment takes place, evicting the poor—who are the easiest for them to move—but leaving the land grabbed by politically and economically powerful agencies and people intact. It is much easier to relocate fishermen than to move the dump out of the marsh and to figure out what to do with 36 lakh cubic metres of garbage, for instance.

Environmentalist and nature-educator Yuvan Aves, along with a passionate team of young naturalists, conducts a series of nature workshops for students at the marsh, helping them identify biodiversity, as well as talk to the people who live near the marsh, such as the fishers and farmers, about the challenges they face. He points out that resettlement, a common 'solution' proposed for restoration, brings its own challenges, as the poor and those with the most insecure livelihoods are always the first to be displaced.

Writer, researcher and environmental activist Nityanand Jayaraman, who has worked in Chennai for decades, also highlights issues of environmental justice, often ignored in discussions of urban restoration. Jayaraman points to the fact that wetlands were locally designated as *poromboke*, i.e., as commons reserved for community and public use. In an article written in 2017, termed 'Disaster by Design', Jayaraman argues that 'it is a fact that the poromboke is the backbone of our economy'. Yet, he says, it is an equally unfortunate reality that in colloquial Tamil, the word has acquired a negative connotation, used to call someone a 'waste fellow', or to refer to a place that is of no use to anyone.

The musician and writer T.M. Krishna sang on the poromboke, drawing public attention to the importance of Chennai's ecosystems. K. Saravanan, a member of the fishing community, is mapping the coastal commons with its creeks, organizing local fishing communities to save their villages. Activists like Saravanan and Jayaraman, public intellectuals like Krishna and many others working in Chennai have also pointed out the futility of demolishing slums to restore rivers, while the buildings of the rich and influential remain in place.

Today, Chennai seems keen to invest in disaster management guidelines and early warning systems. Essential though these are, in an era of climate change, they will go only so far. If coastal cities like Chennai must break out of their seemingly endless cycle of floods and drought, they need to go back to the basics. There can be no substitute for reviving degraded rivers, lakes, ponds and tanks, and protecting them from future degradation. But alongside this, the city needs to be appreciated as a living ecosystem, with multiple interconnected components. Unless public transport improves, the road networks will continue to be expanded, encroaching further into the rivers and canals. Unless Chennai begins to segregate, recycle, reuse and reduce its garbage, the Pallikaranai marsh can never be completely restored.

Rachel Carson, one of the most influential environmentalists of the 1960s, challenged the notion that humans could master nature. She said: 'Man is a part of nature, and his war against nature is inevitably a war against himself.'

By destroying our urban water bodies, we are waging a war against ourselves. This is a war that we cannot afford to continue.

The Apostle's Waters

St Thomas Mount, locally known as Parangimalai (the hill of the *firangi* [foreigner]), is a small hillock situated close to the Chennai International Airport. This hill is where St Thomas, the apostle of Jesus, is believed to have taken refuge and was later martyred. During the early colonial period, water from the St Thomas Mount was filled in cisterns and taken to homes. The water source here has an interesting legend associated with it. Alexander Hamilton writes in *A New Account of the East Indies Being the Observations and Remarks of Capt. Alexander Hamilton from the Year 1688–1723* (Volume 1):

> There is a little dry rock on the land, within it, called the Little Mount, where the apostle designed to have hid himself till the fury of the pagan priests, his persecutors had blown over. There was a convenient cave in that rock for his purpose, but not one drop of water to drink, so St Thomas cleft the rock with his hand and commanded water to come into the cliff, which command it readily obeyed; and ever since there is water in that cliff, both sweet and clear.

FIFTEEN

BONE SWALLOWERS, CORPSE-EATING TURTLES AND CROCODILES IN THE CITY

'*Machhli jal ki rani hai.*'
As the old Hindi rhyme for children tells us, the fish is the queen of the waters. Freshwater constitutes a vanishingly tiny fraction of the world's total area, less than 0.01 per cent. But don't let this fool you. These habitats are incredibly rich in biodiversity and support over 1,25,000 described species. Ecologists categorize the biodiversity in aquatic environments into five groups: from neuston or pleuston, species which live on the surface of water bodies, to periphyton, plankton, nekton, ending with the benthos, species that live at the bottom of the sea or lake. Limnology,

the science that deals with the study of inland waters and their connection with the atmosphere, land and oceans through the transfer of energy and materials, helps us understand why freshwater ecosystems are so important for human survival and also to find better ways of managing them.

However, you don't need to understand inland waters at this level of scientific detail to understand their importance in our daily lives. Look closely at a lake or a river, even one that is polluted, and you will see snails at the edges, clams on the lake bed and water striders on the surface. Swarms of dragonflies and damselflies play in the air, their iridescent wings dazzling in the sunlight as butterflies flit between the grasses and reeds along the banks.

'I Contain Multitudes,' said the famous poet Walt Whitman in his 1855 poem 'Song of Myself'. Whitman was speaking of human complexity but he could as well have been speaking from the perspective of a freshwater ecosystem. Many species lie hidden in the water, lurking between the vegetation at its fringe. One of the largest insects found in freshwater is the water strider *Gigantometra gigas*, found in the waters of Vietnam. While the body itself is short, less than 1.5 in., its legs can spread out to cover an area of more than 10 in.

However, even this does not match the size of the world's largest water insect, a carnivorous giant water bug from Brazil and Venezuela, that can grow up to 4.5 in. The bug has the painful local name of toe-biter. It catches unwary swimmers by taking a nip out of their digits. It can also take on and eat creatures ten times its size, including frogs, turtles and snakes, by paralysing them, injecting them with enzymes that break down their tissue and then ingesting the liquefied meat.

We do not have to worry about the toe-biter in Indian waters. We are more likely to spot the Indian black turtle, easily recognized by its hard, blackish shell. In Hindi, it is known as the *talao kachua* (pond turtle). In Tamil, it is more insightfully called the *pee amai* (urinating turtle). If you pick the turtle up, watch out, it is likely to urinate on you! Another common turtle, the Indian flapshell, has a soft shell in shades of olive brown to grey green. Unlike the urinating turtle, the Indian flapshell has no reason to feel self-conscious about its name, being called *sundari* (meaning 'beautiful') in Hindi and *pal amai* (milk turtle) in Tamil.

Frogs and snakes are other common sights at the waterfront. The Asiatic water snake, also known as the checkered keelback, suns itself on the banks of a river, gliding into the water when it spots an unwelcome visitor. In the lake located in the botanical gardens of Lal Bagh, Bengaluru, we have seen as many as eight keelbacks, some lying in the water and others basking in the sun on the stone embankment.

The insects, reptiles, amphibians and fish constitute an interdependent food web. The frogs eat the insects, and in turn supply a buffet meal for water birds—of course, some insects like the toe-biter eat frogs too! Birds are some of the most documented residents of urban water bodies. Avid birdwatchers make frequent trips to their local water bodies, recording the migrant and resident birds they see. Many birdwatchers who once wrote their bird lists down in a notebook now take their smartphones with them. They upload the details of the birds they see on to apps such as eBird, the online database of bird observations housed at the Cornell Lab of Ornithology. Researchers then use this data to analyse changes in bird diversity.

A recent study we conducted, in collaboration with other scientists at Azim Premji University and Nature Conservation Foundation, used eBird data from the lakes of Bengaluru. We found that the resident bird species which make their homes in lakes in the city around the year are thriving and have even increased in number recently—possibly due to the restoration of several lakes in the past few years. In contrast, migrant bird species, which come from far-flung parts of the world, have declined over time. Because thousands of visitors to these lakes have painstakingly documented their observations on their phone, researchers can get a better understanding of changes in bird diversity over time, helping us better manage the biodiversity in the city.

No such apps exist for other taxa, at least not ones as popular as eBird. In the popular imagination, we do not think of reptiles as important components of a lake, despite the ubiquitous presence of turtles. Unlike birds, they are usually silent, and being well camouflaged in the greenery and water, they are hard to spot. But larger reptiles, such as crocodiles, also make their presence felt from time to time. In Vadodara, Gujarat, the floods of April 2019 brought the waters of the Viswamitra River into homes, offices and roads. Along with the waters came unusual, unwelcome visitors. A news channel described the event thus: '*Shehar main magarmacch. Sadak par magarmacch. Jahan jab nazar paden magarmacch hi magarmacch. Yeh tasweere kisi jangal ki nahin hain, janaab yeh Vadodara hain!*' [Crocodiles in the city. Crocodiles on the road. Wherever the eye falls, we see crocodiles and more crocodiles. This picture is not from any jungle, but sir, from Vadodara!]

The crocodiles that entered Vadodara city were the mugger or marsh crocodiles, one of the two species of crocodiles that inhabit freshwater systems in India. The mugger can grow quite large, as much as 5 m in length. Once endangered, the fortunes of the mugger have been revived by successful conservation and breeding programmes in India, and the species is now found in relatively large numbers in India's lakes and rivers, many of them close to cities.

Herpetologists—zoologists who study reptiles and amphibians—have been tracking the population of muggers in the Viswamitra for a long while. Muggers were first seen in Vadodara in 1985. Scientists suspect that the reptiles came from the Sayaji Sarovar, a reservoir located downstream of the city. From nine crocodiles in 1993, the numbers steadily increased to 228 in 2012, in a 25-km stretch of the river close to the city. People come from the city to the Bhimnath Bridge to engage in crocodile spotting, even as traffic flows in a steady stream. The sensible among them take photographs and marvel at the size of the reptiles, but a few throw pieces of wood and stones at the peacefully basking or half-submerged reptiles, making them move away. The riparian zone of the Viswamitra is an ideal nesting site for the muggers, who lay eggs along the sandy banks. The river is seasonal, and so is the supply of fish, but there is always some food for them to eat, including birds such as cormorants, herons and water hens and even frogs. But they don't stop at wildlife. Some attack goats, poultry and dogs, and feed on the garbage thrown into the river. During the breeding season, when the crocodiles get more aggressive, they can also attack intrusive humans, sometimes killing them.

In India, humans have lived alongside crocodiles for many decades and human encounters go a long way back. Howard Anderson Musser, an American missionary who travelled across India, describes in his book *Jungle Tales: Adventures in India* (1922) 'huge, bloodthirsty, crocodiles', which were hunted by the British in Kanpur and Allahabad. Seated in boats, the men would shoot the crocodiles from a safe distance. Musser writes:

For British Civil Service people, this is big fun; for the natives, however, it is often fatal, as their part in the merry game is to dive after a dead crocodile when he sinks to the bottom, and there are live crocodiles lurking about down there, and many a poor native has been seized and never come up.

Musser was surprised that the Indians would agree to hunt crocodiles, as they considered the reptile to be sacred. According to him, they even sacrificed their children to crocodiles till the British banned the practice. This may have been an exaggeration. Looking back in time, it is hard to conclude anything with certainty. We do know that crocodiles are revered. Adivasis in Gujarat worship the deity Dev Mogra in the form of a wooden idol of a crocodile. The crocodile is also the sacred *vahana* (mount or vehicle) of the Hindu goddess Khodiyar Mata worshipped in Gujarat and Rajasthan. The crocodiles' sacred status does not always guarantee them protection in cities, either from people who throw stones at them, or from riverfront projects such as on the banks of the Viswamitra, which have destroyed mugger nests and habitats. A fear of crocodiles, especially those living so close to a city, may be justified. The crocodile is, after all, a predator at the top of the food chain. Proximity leads to intensified conflict, and mugger attacks on humans, even if provoked, will lead to retaliation especially when there are injuries and death. But there are other animals who present no threat to humans yet suffer the consequences of human violence.

With its toothy grin, the Ganges River dolphin is an endearing creature. Watching it swim on its side with a flipper trailing the river bottom and spotting its dark shape emerge from the water can be a very thrilling sight. It is one of the most endangered mammals in the world, with only about 5000 remaining in the Indian subcontinent, mainly in the Ganga and Brahmaputra rivers and their tributaries. Because it is a mammal and breathes using lungs, the Ganges River dolphin comes to the surface of the water to breathe every 30–120 seconds. It makes a peculiar sound when it breathes, because of which it is also known as 'Susu'.

These dolphins have eyes that are tiny, the size of pinholes, lacking a crystalline eye lens. As a result, they are practically blind. Scientists speculate that because river waters are muddy, making it very difficult for dolphins to see underwater with their eyes, they lost the use of their vision. They evolved to use sonar, or echolocation, to guide their movement underwater. Sadly, this evolutionary innovation of the dolphin did not account for river traffic. The rivers are now filled with boats and trawlers which create underwater noise that interferes with dolphin echolocation. Unable to find their way, they often blunder into boat propellers while feeding, meeting a tragic death.

The presence of freshwater dolphins is an indicator of the health of the river system. But dolphins are highly endangered across the world. There are only four freshwater dolphin species, of which one, the baiji or Yangtze River dolphin, is believed to be extinct. Dolphins are disappearing because of the direct result of pollution, construction of dams and barrages, accidental killing by trawlers and poaching. Their fat, or blubber, is used as fish bait, and oil for medicines, and also as an aphrodisiac.

The dolphin has a long history in India. In his fifth pillar edict, as far back as the third century CE, Emperor Ashoka forbade the killing of the *Gangapuputake* (dolphin). The *Baburnama,* the memoirs of the Mughal emperor Babur, in 1598, describes the *khokk aabi* (water hog or dolphin) in the Ganga, a river in which the emperor liked to swim. In 2009, the Government of India declared the Ganges River dolphin 'India's National Aquatic Animal', but the future survival of the dolphin, especially in India's cities, is in question. It was last seen in Delhi in 1967, when a dead dolphin was fished out from the Yamuna. One of the few cities where it is still found is Patna, along the stretch of water where the rivers Gandak and Ganga meet, and in the waters of the Brahmaputra around Guwahati.

The Ganga may be the most sacred river for Hindus but agricultural run-off, industrial effluents and domestic sewage from different points along its route have polluted the river. Necrotic pollutants, burnt and semi-burnt human corpses taken to the banks of the Ganga for cremation also impact river quality. The dead are cremated in funeral pyres that line the ghats along the water. When the Indian government launched the Ganga Action Plan in 1986, one of the proposals was to release flesh-eating turtles to dispose of the floating carcasses. Just as outsourcing began in India in the 1990s, the cleaning of the rivers was also outsourced, this time to the reptiles.

Two species of turtles—the Indian softshell turtle and the Indian flapshell turtle—were bred in the Sarnath Turtle Centre outside Varanasi, the Kukrail Gharial Rehabilitation Centre in Lucknow and other places and released into the Ganga and Gomti rivers at Varanasi and Lucknow. A staggering 41,000 turtles are estimated to have been released from the Sarnath Turtle Centre alone between 1989 and 2019. To protect the turtles and help them get on with their work, a stretch of the river was also designated as the Varanasi Turtle Sanctuary. Estimates say that ten adult turtles can dispose of an adult human body in two or three days. The turtles can give bathers a nasty bite but generally avoided live people, preferring to feast on the dead corpses floating on the water.

This experiment seemed promising but eventually fizzled out. It is not entirely clear why the turtles disappeared. One conjecture is that local people found the flesh of these flesh-eating turtles to be a delicacy, and the meat made its way to dinner tables. Some speculated that the turtles became a part of the illegal wildlife trade, being transported to China. The Wild Life (Protection) Act of 1972 protects all species of turtles in India by law. But neither the writ of law nor the designation of sanctuaries in Varanasi helped protect the turtles that worked so hard to clean the Ganga.

Douglas Adams, who wrote the cult classic book *The Hitchhiker's Guide to the Galaxy* also wrote a poignant book *Last Chance to See* along with the photographer Mark Carwardine. The book chronicles their journeys to different parts of the world to see some of the most endangered animals including the Yangtze River dolphin, now believed to be extinct. In the book, they ask why we should care about the loss of non-human species. Because, as they say, 'the world would be a poorer, darker, lonelier place without them'.

Our urban water bodies and their surroundings would indeed be a poorer place too without the sound of the Susu, the sight of the basking turtles or the gentle ripples of a gliding mugger.

The Women Who Protect the Hargila

The hargila has neither the beauty of the peacock nor the grace of a peregrine falcon. It sings no song like the koel, nor builds exquisite nests like the baya weaver. It is shunned as a bad omen, derogatorily called the hargila ('bone swallower' in Assamese). The greater adjutant stork's very gait is strange—its common name in English is derived from its stiff adjutant-like walk, that reminds one of a military officer.

It is a massive bird, with a wingspan of 8 ft and a height of 5 ft. Found in wetlands, it hunts for rats, snakes and fish. But the bird is also a scavenger, and for this very reason was much sought after in colonial Calcutta as it kept the city clean and prevented the spread of disease. So useful was the bird as a scavenger, that anyone who caused injury or death of the stork was heavily fined. But this very same bird that the British thought was disgusting to look at for its feathers that grow as a tuft under its tail was also highly prized for another reason. These were used in *marabout*, a trimming of down feathers on clothes popular in Victorian fashion of the 1800s.

Once abundant and wide ranging in South Asia, the adjutant stork is now highly endangered, with a population below 1200. The Zoological Society of London categorizes it as an Evolutionarily Distinct and Globally Endangered (EDGE) species, close to extinction. One of the few sites where it is found in large numbers is the Deepor *beel*, a large wetland in Guwahati next to the city's landfill, where unsegregated waste of plastic, glass and metal, and even biomedical waste is dumped. Amid this toxic garbage, the storks (whose numbers, according to a survey in 2018 stood at 220) nest and feed. In 2017, twenty-five storks were found dead, probably because of the toxins they ingested at the waste site. A heart-rending video of a disoriented stork trying to find its way amid honking traffic in Guwahati's crowded streets did the rounds on social media in 2017. The stork flew away unhurt, but the incident was a stark testimony to the various dangers faced by water birds (and animals) in our cities. Because the stork is believed to be ugly and an ill omen, unlike the charismatic and gentle Ganges River dolphin, few people seem to be interested in battling for its protection either.

But beauty lies in the eye of the beholder. The hargila has a staunch friend and protector in Purnima Barman, who thinks it is the most beautiful bird in the world. A one-woman campaign by this extraordinary fighter seeking to protect this reviled bird has now attracted an army of women. Using a variety of strategies, Purnima has set up a range of innovative programmes to raise awareness of the threats to the hargila—mobilizing schoolchildren, featuring plays with theatre groups, creating stories around the stork, weaving hargila motifs into the traditional Assamese *gamcha*s (traditional scarves) and organizing baby showers for newborn chicks. She is now called the *hargila baideu*, the sister of the hargila. She has won several awards for her work, including the prestigious Whitley Award for conservation in 2017. Working alongside her is the Hargila Army—400 women from Kamrup district of Assam, who work on awareness programmes, engage with the forest department staff in conservation efforts, protect trees with nesting birds and help rehabilitate injured birds. The ugly hargila, once reviled as a bird of ill omen, can breathe easier today, with a fiercely protective sister and an army of women by its side.

'**I** am not the Minister of Water Resources but the minister of water conflicts,'

This was said by India's former Minister of Water Resources, Priya Ranjan Dasmunsi, in 2005.

The minister was speaking about disputes over water, an increasingly common feature in India and across the world. Forty per cent of the world now suffers from scarcity of water. Water shortages impact incomes, exacerbate conflict and increase armed interstate disputes. Drought has even led to extended civil war in countries like Sudan, Somalia, Ethiopia and Yemen.

Conflicts over water were also common during historic times. During the Maratha–Mysore Wars of the eighteenth century, for example, when Maratha armies battled with Hyder Ali and Tipu Sultan for control of the Mysore kingdom, wells and tanks were routinely poisoned, breached or destroyed by each side to deprive the opposing troops of water.

In a world increasingly impacted by climate change and human misuse, we often worry about whether there is enough water for everyone. To manage water scarcity, we need to first be able to measure and monitor it, to track who is being affected over time. This may seem like a straightforward task, but it is mired in complexity. Scientists have developed several indices to measure water scarcity, each with its own advantages and disadvantages. Perhaps the most widely accepted is the Falkenmark indicator of water stress, defined in 1989 by the Swedish hydrologist Malin Falkenmark. The Falkenmark indicator calculates the amount of renewable freshwater (i.e., usable water) available per person per year for a given region. If this index falls below 1700 cubic metres per person per year, the region is categorized as experiencing water stress. If it falls below 1000 cubic metres per person per year, the region experiences water scarcity, and if it is even lower, below 500 cubic metres per person per year, the region experiences absolute water scarcity.

It is easy to see why the index is popular—it is simple, easy to calculate and can be applied globally. A 2019 analysis shows that the amount of water available per person per year in India has steadily declined since Independence, going from 5177 cubic metres in 1951, to 1545 cubic metres in 2011. One number cannot provide a complete picture of what goes on in a country as diverse as India, of course. Some parts are water-rich, and may suffer from a surfeit of water, while other regions are drought prone. Some regions, such as Chennai and its suburbs, face the problem of flooding in the monsoon and drought in summers. The Falkenmark indicator has its limitations, in that it cannot account for fluctuations throughout the year, or from one part of the country to another.

Yet, there is another challenge of water scarcity that is important to understand in a country as socio-economically complex as India. The availability of water, which should be a universal human right, is in reality a reflection of one's power and status in society. One cannot understand it *just* as a consequence of real or biophysical scarcity. Sociologist Lyla Mehta has worked on the politics of water access in India, Europe and Africa for decades. She demonstrates, through powerful case studies, how water scarcity is socially constructed because of societal norms that people with more power and influence impose on those at the bottom of the social hierarchy.

In an influential paper published in the *Economic & Political Weekly* in 2003, Mehta discusses the issue of water scarcity in the semi-arid regions of Kutch, which is one of the reasons given for the construction of the Sardar Sarovar Dam on the Narmada. Within Kutch, she demonstrates how only certain families face challenges of water scarcity. Wealthy, upper-caste families secure their rights to water by investing in digging wells, while poorer Dalit households are often forbidden from using these wells. Though illegal, this is a problem that Dalit families continue to experience across India, as horrific incidents across the country continue to remind us. How does one calculate the social complexity of water access and exclusion, which leads to violence in many instances, with a single metric? It seems impossible. But that which is not measured is often not inserted into policy targets and plans. And so, how to measure water scarcity continues to be a challenge for hydrologists and sociologists alike.

Other complexities abound. Falkenmark introduced a second globally influential concept in 1995—that of green vs blue water. Green water is the water that precipitates from rainfall, which is held in the soil, drawn up by plant roots, and absorbed by plant growth. Blue water is the water that remains after plants take up the rainfall, flowing from the soil into lakes, rivers and the ocean or percolating inside to recharge groundwater levels. India relies heavily on groundwater to irrigate its crops—we are the world's largest consumer of groundwater, taking out more than the next two countries (China and the USA) combined. Some of the groundwater we extract is replenished by new infiltration from blue water, but much of it is also very old. As we dig borewells to deeper depths, we mine ancient water. In an instant, using fossil-fuel powered pumps, we extract water that collected slowly, over hundreds or thousands of years. This is not sustainable, and the consequences are easy to see. Borewells are running out of water, not just in one part of the country but in many regions.

Fortunately, we do not witness armed battles over water as commonly in India today as during the time of the Maratha–Mysore Wars, though we do see water conflicts of various kinds. Every Indian has the right to water—a right protected by the Indian constitution. Article 21 of the Indian constitution enshrines the Right to Life. This includes the right to clean drinking water, for without water, there can be no question of human life.

But the question gets a little more complex when we move a level higher, speaking not of the rights of an individual but the relative rights of a village, neighbourhood, city or region. When a river originates in a mountain in one state, flows across the plains and valleys of another state, and empties into the sea, forming a delta in a third state, hundreds of kilometres from its source—which of these states has the preferential right to the river's water? Within the state, who gets priority? Is it the communities at its source, the farmers along its route, the cities and towns whose pipes extend to the river, or the fishing communities at the estuary where the river meets the sea? Does the government allot the major share of water to the farmers who feed the country; the industrialists whose money propels the economy, or the cities who lay claim to India's future?

There are no easy answers to such a question. Not for the Union Minister for Water Resources, or for any of us.

Take the Cauvery, for example. The river originates in the springs of Talacauvery, in the Brahmagiri Hills of the Western Ghats, in the state of Karnataka. Do the communities living here have the greatest right to the river, here, where it all begins?

After it emerges from the ground at Talacauvery, the Cauvery traverses 800 km across the Indian peninsula. It covers 320 km in Karnataka and 416 km in Tamil Nadu, with small extensions into Kerala and Puducherry, eventually emptying into the Bay of Bengal at Poompuhar on Tamil Nadu's coast. Shouldn't the fishing communities in the estuary, at the end of the river, have the right to its waters?

Sixty per cent of the water of the river is used for irrigation. The Cauvery delta is called the rice bowl of south India, producing almost half of Tamil Nadu's rice, and feeding millions of people. Rice farmers should have an inalienable right to water, many will argue. Yet the same river also generates hydropower, and supplies cities like Mysuru and Bengaluru with drinking water. Why shouldn't Bengaluru and Mysuru, with their high population densities and contribution to the country's economy, have a right to the river?

The question gets even more complex when we consider geographies and accidents of location that determine attributes of belonging and exclusion. Mysuru is close to the Cauvery, and within its watershed. Bengaluru lies 100 km away from the main river. The state capital needs water to survive, you might argue. But why should a city like Bengaluru receive 540 million litres per day from the Cauvery, in preference to the other communities of fishers and farmers, industries and townships, villages and towns that lie along its course?

Questions of rights and privileges become very complex in such situations. As do the legal arguments and political debates.

Beginning in British colonial times, there has been a long history of interstate disputes over rights to the water from rivers. Tribunals set up to resolve disputes around water-sharing in many rivers in India—Narmada, Godavari, Krishna, Ravi, Beas, Mahadayi, Mahanadi, Vamsadhara and Cauvery—have met with varying success. The Cauvery Water Disputes Tribunal (CWDT) has had a turbulent history, sparking two especially violent riots in Bengaluru. In December 1991, when the CWDT was directed by the Supreme Court to release 2,50,000 million cubic feet to Tamil Nadu, Bengaluru erupted in violence. Property was destroyed, and normal life disrupted for several days. In September 2016, another dispute led to the burning of thirty-five buses, injuring several people.

In 2016, the Supreme Court gave its final orders on water sharing, ruling that more water was to go to Karnataka. A newspaper headline in the *Times of India* from this time announced that 'Karnataka gets more Cauvery water, thanks to Bengaluru'.

Why such a title? The Tribunal had earlier decided that Bengaluru would only receive a certain fixed amount of water, since only one-third of the city fell within the geographic boundaries of the Cauvery Basin watershed. The Supreme Court however ruled that the geographic location of Bengaluru could not be the deciding factor. Other factors needed to be considered while apportioning a fair share of water to the megacity. Elaborating on its decision, the court said:

> The city of Bengaluru . . . is incomparable in many ways not only to other urban areas in the State, but also beyond. The requirements of its dependent population as a whole for drinking and other domestic purposes, therefore, cannot justifiably, in the prevailing circumstances, be truncated to their prejudice only for consideration of its physical location in the context of the river basin.

And thus, a city's thirst may sometimes seem to outweigh the needs of others.

Another contentious dispute is over the waters of the Krishna River, which originates near Mahabaleshwar in the Western Ghats. After covering a distance of about 1400 km across three states—Maharashtra, Karnataka and Andhra Pradesh—the Krishna drains into the Bay of Bengal. One of the questions before the Krishna Water Disputes Tribunal was similar to the question that the Cauvery Tribunal considered. How much water should each state give to Chennai to meet its drinking water needs? This may seem to be a peculiar question if one looks at the geographic location of Chennai. The city is the capital of Tamil Nadu, a state through which the Krishna River does not even pass! But the megacity of Chennai, lacking its own perennial water source, needed an alternative. The states involved worked out an agreement. Andhra Pradesh would provide water to Chennai from the Krishna River, while Tamil Nadu would fund the creation of a canal network that also supplied water to farmers from arid parts of Andhra Pradesh.

Courts and tribunals debate and decide on the fine points of interstate water sharing disputes, but who decides how water is distributed within a city?

The National Water Policy of 1987, updated in 2002 and 2012, accords a higher priority to drinking water, placing it above other needs. Yet we often come across a long line of people (usually women) carrying pots and buckets, queueing up to collect water from a tanker or a pump in a city. Is this what our urban future will look like? With some of us drinking bottled water from Himalayan springs, or spring water from the French Alps, while others lack basic access to clean water to drink, bathe and keep themselves clean?

The National Water Policy of 2012 clearly specifies that 'the principle of equity and social justice must inform use and allocation of water'. This is well stated but hard to carry out, in part because of the speed at which cities are growing today. Many pockets of informal settlements, urban slums and shanties lack access to a reliable water supply. In a stressful urban environment, where water supplies are low, and the money to purchase expensive water from private sources even more scarce, quarrels around public taps are an unfortunately common sight.

As we move outwards from the city core to its periphery, water supply moves from the formal to the informal. Piped water supply is unable to keep pace with the spread of the city. In the periphery of the city, both the rich and the poor must deal with scarcity as best as they can. The rich invest in digging borewells and sourcing water from tankers. This further depletes the groundwater supply. The poor purchase water in smaller quantities, in pots and buckets, often paying much more in relative amounts compared to the middle class. Caste plays a role in shaping water access, and so does gender—women from underprivileged caste groups face the brunt of water scarcity, whether in our villages or cities. Driven to desperation, those who lack water may resort to using dirty water to meet their needs, washing clothes and vessels in ponds or lakes that are contaminated with garbage and sewage. This is the reality of India's cities, where lofty principles of social justice and equity seem to be at odds with the reality of urban life amidst water scarcity.

Water tankers may seem to be everyday saviours, supplying water across the city, even during times of drought and scarcity. But who operates these tankers, and where do they get their water from? Many researchers and activists term them the 'water-tanker mafia', because of their tight control on the profitable networks that supply water across the city for a price. In Delhi, tankers are filled with water from borewells, dug on the banks of the Yamuna in the dead of the night. An intricate network of suppliers takes water to the city—including the person who extracts the water (often illegally) from the Yamuna riverbank, the tanker owner and the intermediary who connects customers to suppliers. The capital city's malls, hotels and high-rise apartments seem to have an endless thirst for water, and tankers operate to fill this need. Government officials in water departments sometimes speak candidly in interviews, claiming that the mafia does not allow the expansion of formal, low-cost water supply networks. The mafia seems to develop in cahoots with local leaders, they say.

Thus, we see that the core areas of the city, and the houses of the wealthy, are prioritized by the municipality and by planners, over the peripheral areas and the urban poor. The city claims priority over the river, wresting water from rural users. But when it comes to sharing within the city, the poor always seem to be left out.

In 2019, when Chennai was going through a water crisis, a spat over water led to one neighbour stabbing the other. This is an extreme consequence of a water conflict, but it warns us of the possibility of similar incidents in the future, if water scarcity persists. Fault lines are beginning to appear, showing the breakdown of trust within communities, with some societies installing CCTVs to detect the theft of water, and people using social networks to influence the distribution of water to one home instead of another.

Yet, all may not be lost. Water is a renewable resource, and if cities can harvest rainwater, recharge groundwater, and treat and reuse grey and black water, they may be able to drastically reduce their water usage. Singapore provides a singular example. The island nation-state faced a major challenge of water scarcity in 1965, when it gained independence. Even today, Singapore imports half of its total water supply from Malaysia. Given the small size of the island, its capacity to harvest rainwater is limited. Yet the country has made substantial progress in water management by treating and reusing its wastewater and reducing wastage. With these changes, Singapore's per capita water consumption is just 158 litres per person each day, quite low compared to other industrialized countries such as the USA whose usage is more than double (380 litres per capita). Similarly, residential layouts such as Rainbow Drive and Zed Earth at the periphery of Bengaluru have drastically reduced their dependence on water tankers, by treating and reusing wastewater, aggressively implementing rainwater harvesting, recharging rainwater into the ground and reducing their overall water consumption.

The World Bank warns that 40 per cent of the world is currently facing a water crisis and this number will only increase in the future. If this happens, water wars will no longer be the storyline of apocalyptic films— they will become the everyday reality of our cities.

As the saying goes, we need to wake up and smell the coffee (or the polluted water)—to act!

SEVENTEEN

BENGALURU: LANDLOCKED CITY OF TANKS AND LAKES

On a map of Bengaluru, you might notice several place names that end in *'sandra'*—Dommasandra, Singasandra, Jakkasandra, Lakkasandra, Bhoopasandra, Hongasandra, Mallasandra, Junnasandra.

'Sandra' is a shortened version of the Sanskrit word *samudra*. While we think of 'samudra' as synonymous with the sea, it refers to any large body of water. How did the landlocked city of Bengaluru have so many locations whose names indicate the presence of *samudra*s which now lack all traces of water? This is the unfortunate story of Bengaluru, once called the *kalyana nagara*, city of reservoirs or tanks, whose water bodies have systematically been erased from many parts of the city.

Bengaluru seems like an odd location for a city. Located in the rain shadow of the Western Ghats, it is semi-arid, with relatively low levels of rainfall. The city is not near a coast or delta and has no perennial rivers or lakes. And yet, the presence of megalithic stone tombs shows us that people lived in this landscape at least about 3000 years ago. We have no idea how these ancient people lived, or what they did to get water. But hundreds of epigraphs, inscriptions on stones and copper plates from about the sixth century onwards give us some information about how new settlements were developed and what they did for water.

Before it became a market town and urban centre, the landscape that comprises Bengaluru city today was a collection of thriving villages that made a living from agriculture, cultivating dry millets and irrigated paddy, growing orchards of fruits, flowers and coconuts, livestock herding and fishing—occupations that depended on reliable water supply.

Several dynasties ruled the region of Bengaluru, with the Gangas, Cholas, Hoysalas and the Vijayanagara kings prominent amongst them. These early settlers took advantage of the undulating topography of the land, and the ridges that divide the region into two river basins. The eastern basin drains into the Arkavathy, a tributary of the Cauvery River, and the western basin connects to the Dakshina Pinakini River, known as the Pennaiyar once it reaches Tamil Nadu. The early inhabitants of Bengaluru further sculpted the topography of this undulating terrain to make the land inhabitable, carving out tanks, which we now call lakes, from natural depressions. As much of the rainwater as possible was harvested in these tanks and the porous ground around them. Later rulers—the Kempegowda rulers, the Maratha warrior-king Shahji Bhonsle, the Wodeyar kings of Mysuru, the Mughals who ruled the city for a brief three-year period, Hyder Ali and Tipu Sultan, and the British—all knew the importance of water, maintaining the lakes and ponds, and creating new ones.

Near the old airport of Bengaluru, a stone inscription from 1307 CE, in the village of Vibhutipura, describes how one of these tanks was created in some detail, when people 'cleared the jungle in the tract of land adjoining Peru-Erumur, levelled the ground, built a village, constructed a tank by removing the sand, and named the village Vacchidevarapiram'. When lakes at a higher elevation filled with rainwater, the excess water was directed into channels called *kaluve*s, which transferred water into the next lake lower down. In this manner, lakes were connected into networks, with the land in between forming richly productive wetlands, used for cultivation, cattle grazing and fishing. The larger lakes were called *kere*s, and used for agriculture, while smaller reservoirs, including *kunte*s and *kalyani*s, were used to wash cattle and clothes, and to provide drinking water, along with large open wells that surrounded the lakes.

Over a thousand of these water bodies, ranging in size from large to small, covered the surface of this region. Some were said to have been constructed by wealthy patrons such as local chieftains and others by courtesans, but they were built by the (often coerced) physical labour of poorer communities, especially Dalits. While the contributions of the patrons were recorded in inscriptions and later in books on the city, the toil of the latter often went unmentioned. During our research on lakes in the city, we have often heard deeply disturbing stories of human sacrifice associated with the building of the lake, usually of women and children. Sometimes, we have been asked to walk cautiously and talk softly while approaching a lake bund at sunset, so as not to disturb the spirits of a mother and child who were sacrificed and buried at the lake. Poignant songs sung by elderly women narrate stories of a daughter-in-law of the local leader who sacrificed herself at the lake after a prolonged period of drought or flooding. Each lake has a shrine to the local lake and village goddesses. Festivals celebrated at these temples commemorate times of plenty, when the lake overflows with water.

Once focal social, cultural and economic points for the local community, these lakes are now threatened. Bellandur, Bengaluru's largest lake, has made headlines across the world for its toxic foam that catches fire. Much of the landscape between the two largest lakes of Bengaluru—Bellandur and Varthur—has been built on, with the productive wetlands and grasslands converted to concrete buildings. Because the soil is covered with impermeable concrete, rainwater is unable to percolate into the ground, resulting in widespread flooding during the monsoon. These lakes are at the periphery of the city, but the central areas are hardly better planned, at least in terms of their water systems. Almost all the lakes in the older, central parts of the city have been lost, filled in and built over.

Sampangi Lake is now the Sri Kanteerava Stadium, the Dharmambudhi Kere holds the city's central bus station and Shoolay Lake contains the Garuda Mall, to name just a few. Ulsoor Lake, Sankey Tank and Agara Lake are some of the very few water bodies remaining in the heart of the city, and even these are quite polluted. Our research found that between 1885 and 2014, the city centre lost close to sixty lakes and water bodies, and a staggering 1911 open wells—in large part because water began to be supplied from neighbouring lakes in 1896, rendering the storage of water within the city irrelevant for planners. In 1926, water from the Yele Mallappa Shetty Lake, east of Bengaluru, was used as a supplemental source, and another reservoir was constructed in Tippagondanahalli in 1930. Today, much of the city gets its water from the Cauvery River, located at a distance of 100 km.

Meanwhile, planners forgot that lakes were important for flood control and groundwater recharge. They re-imagined these spaces of water as commercially valuable real estate, converting them in a few decades to malls and stadiums, bus stands and layouts. For the British, and later for Indian planners, lakes were also considered cesspools of disease, which bred malaria-carrying mosquitoes, and needed to be filled in to improve the health of the city.

To know where a lake once was, Bengaluru has only to wait for the monsoon. The Kempegowda Bus Stand, the Sri Kanteerava Stadium, the hockey stadium in Akkithimmanahalli in the city and the Asian Games Village in Koramangala, all of which are sites of former lakes, become flooded. With many of the channels that connected these lakes built on, or filled with garbage, water flows on to the road and into offices and homes, throwing parts of the city into chaos for weeks. Meanwhile, planners, politicians, real estate honchos and corporate giants begin a ritual blame game, each tossing the ball of responsibility to the other. In the end nothing changes—only wait, it does! Things get worse, year after year.

With the effects of climate change already being felt, Bengaluru faces an uncertain future of intense floods interspersed with long periods of drought. The city is seeking a new paradigm, a new imagination of how to live with water.

Bengaluru is well known in India for the citizen activism that has led to the revival of many once-degraded lakes, contributing to the physical and mental health of lakhs of people who visit these public spaces to walk, jog and connect with nature in the city. Civic groups and local residents have taken to the streets in protest, engaged in signature campaigns, used social media and filed public interest litigations (PILs) to push for lake restoration. However, many restored lakes disallow traditional users whose ancestors once maintained these lakes, including grazers, fishers and fodder collectors. While the restored lake becomes a location for walking and jogging along well-defined paths, those who depended on the lake for food and livelihoods are alienated.

Signs of hope are beginning to emerge amidst the tales of gloom and doom. Several of the successfully restored lakes in Bengaluru, including Jakkur and Kaikondrahalli, are challenging this paradigm of lake restoration by enclosure. The lake groups here are working with fishers and grazers in innovative ways and lobbying with the municipal government to change their restrictive policies. So are civic organizations like the Environment Support Group, whose landmark PILs have helped provide legal protection for the lakes in the city. Recent efforts at the restoration of kaluves and wells in parts of the city, though very nascent, are also promising.

When Bengaluru was first populated, scattered lakes were first created, and then interconnected to form a living system. So too can these diverse, dispersed initiatives, if they connect and gather force. Then we can see Bengaluru return to being a kalyana nagara, city of lakes, once again.

Origin Myths of Lake Formation

Bellandur, Bengaluru's largest lake, has an origin myth told by many, including the priestess of the Dugalamma Temple on the lake bund. In one version, the location where the lake now exists was once a banana plantation. The goddess, Dugalamma, came in the guise of a pregnant woman (in another version of the tale, an elderly woman) who asked the owner and others for a banana (in another version, for a plantain leaf to use as a plate). When her request met with refusal, the angry Dugalamma invoked a heavy rainfall that submerged the entire plantation, creating the lake that exists today, only stopping the rains when the village built a temple to her. Another origin story tells us that the lake was created in gratitude for an act of kindness, when someone offered water to a thirsty woman (the goddess in disguise). As a blessing, she converted the fallow land into the lake.

These two myths encapsulate the identity of water—a blessing in times of drought and a curse when it floods.

EIGHTEEN

WATER WARRIORS: COMBATING SOLASTALGIA WITH ACTION

Solastalgia, ecological grief, is a feeling that many of us have experienced. Solastalgia is derived from the Latin root word *sōlācium* (comfort) and the Greek root word *algia* (distress). The term describes the feelings of sorrow and anxiety about the future that people develop when living in an environment which is deteriorating rapidly around us. First described by the environmental philosopher Glenn Albrecht in his 2005 article, 'Solastalgia: A New Concept in Human Health and Identity', the term has become increasingly popular in recent years, when so many people are anxious about the environmental destruction of the world and worry about the growing challenges of climate change.

Solastalgia differs from nostalgia, which describes the emotions of separation and loss we experience after leaving home. Solastalgia describes the loss we feel *while living at the same home* but seeing that the home is crumbling around us. When the two of us walk by a lake in Bengaluru that we once knew well and loved, seeing the once-clean waters choked with stinking trash, and blackened and frothing with toxic foam, ecological grief consumes us. If post-traumatic stress disorder is something we all fear, this is its opposite—pre-traumatic stress disorder, the fear of losing our environmental future that worries us.

Solastalgia is a real fear. But we cannot afford to let it get out of hand. Getting paralysed by anxiety is unhelpful, and deeply damaging, not just to us as individuals but as collectives, as a society. The other side of solastalgia—coping by ignorance, by closing our eyes to the problem while we return to our hectic daily pursuits, or transferring the blame to others citing bad governance, corrupt builders or lack of civic sense—is equally pointless.

Closing our eyes to the problem won't make it go away. Opening our eyes and seeing the scale of the challenge is overwhelming.

How can we deal with solastalgia?

To understand this, we turned to water warriors. Amidst the gloom and doom stories of pollution, degradation, decay and loss, India's cities also hold several committed citizens waging battles to protect the water bodies in their beloved cities. These water warriors wield no swords. But they have an arsenal of resources with which they tackle their self-defined tasks of saving the city's waters. Some go from home to home, spanner in tow, repairing leaky taps and installing water-saving devices. Others build sustainable homes which rely on rainwater, holding open classes to transfer their learning to others. A few intrepid water warriors take up the path of environmental activism, filing PILs and challenging apathetic municipalities to protect wetlands without exacerbating social justice issues. Others prefer the path of environmental education, documenting biodiversity in water systems and sharing these tales of joy and wonder with young children to motivate them to protect their water systems. These water warriors are as diverse as India's urban water bodies themselves. They are the stewards of our city's threatened ecosystems, combating solastalgia and the paralysis it engenders with the forward thrust of environmental action and the optimism it generates.

What inspires these urban water warriors, these environmental stewards, to act in the face of obstacles? Where does their sense of hope come from? With the help of our colleague Sukanya Basu, we spoke to water warriors who live in large metropolitan cities and small towns, involved in protecting nature such as coastal mangroves, wetlands, rivers, ponds and even wells. Their passion and commitment to act for the environment came from a wide range of factors.

In her autobiographical essays, collected in the book *Moments of Being* (1985) Virginia Woolf calls the period of childhood a 'great Cathedral space'. Just as cathedrals, which are imposing, important, unforgettable spaces, childhood is a very formative and unforgettable period in our lives, one that shapes our way of thinking, acting and being especially in terms of our relationship to nature. The same is true of us. Looking back at her childhood days, Seema traces her love for nature to the many hours she and her brother spent climbing trees in the wooded university quarters in which she grew up in Visakhapatnam. For Harini, it is the long walks she took in the parks and tree-lined avenues of Delhi and Bengaluru with her parents, and the walks she took around the restored lakes of Bengaluru with her own daughter when *she* was a young child.

For many of the water warriors we spoke to, childhood memories, experiences and relationships played a seminal role. These experiences established the first connection with nature, that led them in the direction of environmental action in their adult life. This was only possible, they said, because they grew up in environments where they were surrounded by nature. Some of the people we spoke to had grown up in rural villages, in agricultural families, helping with farming, surrounded by animals and plants. Or had the chance to experience rural life in the midst of nature when they visited their grandparents in the holidays, as both of us also did. Others lived in cities but close to a forest, played in a green neighbourhood park, or had a garden in their backyard full of trees and plants. Some of us had teachers who ignited the spark that got us interested in nature, or spent time immersed in books and libraries, seeking to learn more about birds, butterflies and wildlife. These experiences of varied kinds play a role in combating solastalgia today.

It is important to ensure, when engaging with the environment, that nature is not just viewed passively but forms a part of our lived experiences. Throwing a stone into the still waters of a lake and watching it skip on the surface or sighting a bird swooping down into a river to pick up a fish brings back childhood memories, associated with the touch, feel and smell of all things in nature. Climbing trees, chasing butterflies and dragonflies, and watching cattle graze contribute to creating bonds with nature. Many of the people we spoke to described fond memories woven around ponds, streams and rivers—of swimming with friends, plucking lotus blossoms, playing with frogs and taking trips with a grandmother to the river to bring back pots of water. Today, many parents shudder at the thought of children playing in the rain and mud and picking up unknown ailments. But for many of the water warriors we spoke to, the sensory contact of being in water was their strongest, and often fondest, memory.

The Mula–Mutha River provides drinking water to the people of Pune. Shailaja Deshpande has a background in child psychology and had previously worked in the field of inclusive education as a teacher. More recently, with her interest in interior designing, she secured a diploma to pursue her interest and help with her husband's interior design business. But amidst all this, as she says, 'My passion for nature was always there.' As a water warrior, she has now spent many years working to protect and revive the Mula–Mutha River. She says the attachment stems from her childhood spent on the banks of the Krishna River. She remembers going to fetch water from the river with her grandmother. The waters which she swam in as a child, and even drank, are now extremely polluted. She wonders how she can pass on the joy she felt at the river as a child to the next generation. As she says, 'Our body is almost 70 per cent water. Don't we have a responsibility for protecting the river whose water we drink?'

By conducting river walks, a river festival on World Rivers Day and innovative campaigns including one to adopt a stretch of a river, she and her colleagues fostered a strong community working on river protection. They collaborated with local fisherfolk, who lived in the polluted areas, and informed them when untreated sewage was let into the river. The community group then spoke to government authorities, getting them to intervene—a win-win situation for all. She feels especially gratified that the income of the fishermen has increased since they formed the network, cleaning and maintaining the river.

Paras Tyagi had studied physics and public policy from Delhi. He grew concerned about the degradation and disappearance of water bodies as well as trees and parks in his childhood home, an urban village on the outskirts of Delhi. Along with others, he developed an index to rate the environmental condition of water bodies using a scale of zero (low) to seven (excellent). They mapped and assessed 1009 water bodies across Delhi and were horrified to find that none of them had a rating above four. They organized a campaign, getting people from the localities he engaged with to send 11,000 postcards to the Chief Justice of India's office to highlight the state of water in Delhi, and seek judicial intervention.

Paras lives on the land his grandfather once farmed. His childhood memories are of a place full of tall trees and open spaces. Now, the area is covered with concrete, and the farms have disappeared. Modernization of the urban village has provided road connectivity, electricity and the internet, but he still has fond memories of sailing paper boats during the monsoons. Today, in the same urban village where he grew up, Paras has built a connection with the people of the community, encouraging them to protect water bodies.

Not all environmental stewards were fortunate enough to grow up in an environment surrounded by nature, or work with their grandparents on a farm. For some, their connections with nature were shaped later in life, in late youth or adulthood. Many college students connect with nature. It is their time living and studying in large, wooded college campuses, hanging out with friends in animated *adda*s (gatherings), when they become interested in birds, trees and water bodies. One engineer describes a serendipitous walk through the Sanjay Gandhi National Park in Mumbai, that set him on a journey of nature protection. For others, an interest they took up after retirement, cultivating home gardens, transformed into a passion, and they encouraged others in their neighbourhood to take up community gardening in open spaces. Some began to dabble in hobbies such as photography and rowing in the Adyar River, which later took on another aspect of nature protection. They may have started later in life, but as they describe it, their zeal to protect is often even stronger, seeking to compensate for decades of distance from nature-care. These warriors of the environment hail from cities and towns across India—from metros such as Delhi, Hyderabad, Pune, Chennai, Kolkata and Mumbai to smaller towns such as Jabalpur, Dehradun, Guwahati, Coimbatore and Ahmedabad. Their professions are equally varied—homemakers, artists, architects, film-makers, lawyers, software professionals, engineers, retired teachers and bankers.

The East Kolkata Wetlands (EKW) owe much to the involvement of many civic activists including the indefatigable Dhrubajyoti Ghosh. Nobina Gupta, another steward of the EKW, initially visited the wetlands as an artist, observing and enjoying nature. Nobina is the founder of Disappearing Dialogues Collective, which works with local communities, especially children, living in the wetlands landscape. She has been a witness to the disappearing mosaic of the blue and green landscape of the wetlands as they were being converted to commercial and residential spaces. She became worried that the idyllic *baganbari* (holiday destination) was slowly degrading in front of her very eyes. She also saw how this excessive urban growth affected the fishers and farmers who earned their income from the wetlands.

Nobina engages with children from government schools who also live in the wetland, getting them to write stories of their life (called *amader kotha*) and then extending these stories to highlight the importance of the wetlands in which they live. The children share their knowledge about the wetland, observing and sketching by observing the biodiversity and landscape, collecting stories related to their areas from their families including grandparents, and documenting traditional recipes. Art, culture and food are catalysts that generate interest from children who live in the wetlands, whose parents' livelihoods depend on the wetlands. She works with artists to conduct drawing and painting workshops for children and has curated installations featuring the landscape of the wetlands.

One of these is a Pattachitra (traditional painting done on cloth base) panel showing the wetlands landscape. Through these efforts, she seeks to create future stewards of the wetland.

Kalpana Ramesh is an architect and designer who spent her childhood in Bengaluru, the Garden City, in a house with a front yard covered with rose bushes in bloom. She now lives in Hyderabad with an equally enthusiastic nature-loving spouse. However, it was not nature-watching but an awareness of their dependence on water supply from tankers in Hyderabad that got her interested in rainwater harvesting. Concerned about the quality and cost of tanker water, she realized that this was not a sustainable water source. As she describes it, she did not begin this work because of a specific emotional connection to water. Instead, it was gardening and tree planting that she was interested in. She found it prohibitively expensive to foster this interest through the purchase of tanker water, which was also unreliable and difficult to procure. This experience got her started in investing time and energy in understanding rooftop rainwater harvesting. While it was not an easy task, once they implemented rainwater harvesting, she and her husband found their dependence on tankers reduce considerably.

She remembers living through the summer of 2016, a time when Hyderabad witnessed massive water scarcity, and when workplaces started to limit the water used in their offices. By then, Kalpana had extended her water harvesting efforts from their home to her community. The drought of 2016 prompted her to ask: what about the city? She dove into the history of the city's water system, looking at the historical tanks, stepwells and lakes, to understand how the city had carefully planned its water systems earlier. She saw how disconnected modern urban planning had become from water. Focusing her efforts on a lake near her house that was in a degraded and polluted state, Kalpana worked with local residents and government officials to raise funds for rejuvenating the lake. After the success of this effort, she has now decided to take on larger challenges of looking at water conservation at the city level.

Today, we have Instagram influencers whose endorsements are a rage among children and adults. For the warriors, influencers were often closer home—parents, grandparents and teachers. For some, the care and connect in adulthood towards nature is linked to gardening with parents and grandparents, walks in nature, sports engaged in wooded playgrounds, and endless cups of tea drunk in college *chai kade*s (tea shops) under a large tree or by the side of a lake. Warrior moms take action against issues like air pollution, concerned about the future world their children will inhabit. Senior citizens seek to create environmental memories for children just like the ones they had been lucky to have.

Solastalgia is not just a self-centred emotion. Yes, we worry about our individual futures, but we are also consumed with anxiety about our friends, communities and fellow beings, and the other species who inhabit our world.

Combating solastalgia through environmental stewardship, caring for the environment, can be extraordinarily enriching and rewarding, often ending up becoming a lifelong journey of passion. The beginning can be simple, easy—it need not be large or overwhelming. Anyone of us can be an environmental steward by making a start, selecting any mode of engagement that appeals to us most. We can sign a petition or attend a protest; participate in a trash clean-up or start a composting club; photograph birds, start a blog, or form a children's group to conduct bird walks around local lakes and rivers; participate in ward committee meetings, and even run for office. Whichever one of these we seek to begin with, we should not overthink it before we begin. Diving in, and even making mistakes along the way, we learn and reflect on the strategies that worked and failed for us, gradually broadening the circles within which we engage and view the world, to nurture our connection with nature. Immersing ourselves in the environment, through action, is the best cure for solastalgia. We owe it to ourselves, and to future generations who will inherit our world.

NINETEEN

DEALING WITH CLIMATE CHANGE

Climate change has made it to the daily news nowadays. Yet despite knowing of the threats of global warming in 1896, we have been slow to recognize the threat that it poses to our lives and future—and even slower to act.

How will climate change affect urban India's water security? And why should we care?

'We forget that the water cycle and the life cycle are one,' said Jacques-Yves Cousteau, the legendary oceanographer and film-maker, known for his stunning documentaries of underwater life. Cousteau was speaking of the water cycle, which children learn about in school. Under the heat of the sun, water evaporates from the surface of the earth, rising into the atmosphere. As it rises, water vapour cools and condenses into droplets of rain, falling on to the earth and collecting in lakes, rivers and oceans. As the sun's rays fall on the earth's surface, the process repeats again and again in an endless cycle.

Climate change speeds up the water cycle in some places and slows it in others. As the world warms, evaporation will increase in regions of land that contain large bodies of water, such as river basins and glaciers. Such locations will experience more rainfall, in intense bursts. But parts of the world that are already dry, off the path of the monsoon winds, will have less rainfall, and become more drought prone.

India is one of the countries that will constitute 'ground zero' for climate change. Now the world's most populous country, India features in the list of top ten countries most affected by climate change, with a billion-plus people at risk. The Global Climate Risk Index 2021 estimates that in one single year, 2019, India suffered economic losses of $68,812 million (in purchasing power parity, i.e., after accounting for the differences in the purchasing power of the US dollar in the USA and the rupee in India).

A growing economy like India can ill-afford such damage. And yet, this is just a glimpse of the impact climate change will have on our country's economy in the decades to come.

Water plays a crucial role in enabling India's economic growth. The Indian stock market tracks the monsoon closely, rising when the meteorological department predicts a good monsoon, and falling when there is news of delay, or a below-average rainfall. Sixty per cent of the country's population depends on agriculture in some form or the other, and twenty per cent of the gross GDP comes from agriculture.

India's farmers depend on the monsoon. Globally, 70 per cent of freshwater extracted is used for agriculture. In India, this proportion is even higher, with 80 per cent of water being used to grow crops. The glaciers and snow-covered peaks of the Himalayas feed the three major perennial river systems—Ganga, Brahmaputra and Indus—which supply freshwater to most of northern India. These three rivers collectively account for about half of India's total surface water supply. Rainfall also contributes to the water in these rivers, and a bad monsoon reduces the availability of surface water in north and east India. In central and south India, where most rivers like the Narmada, Mahanadi, Cauvery and Krishna are seasonal and largely rainfed, a bad monsoon can be devastating for farmers, especially in the south-east where rainfall is already low. The situation is even worse in the arid regions of Rajasthan, whose terrain is marked by seasonal rivers and streams.

Despite some exceptional years of high rainfall, there seems to be a shift in the timing of the monsoons. Climate change has also worsened our capacity to predict the arrival and departure of the erratic Indian monsoon, making an already unpredictable situation even more uncertain. Agriculture needs the right amount of water, at the right time. Farmers need to know when to plant and when to harvest. As the timing and amount of rainfall becomes increasingly unpredictable, agriculture becomes a gamble—crops fail and pest attacks increase.

The South Asian monsoon is extraordinarily complex and has always posed a challenge for climate and weather models. Overall, the rains appear to be making a later entry into the Indian subcontinent and withdrawing earlier than they used to. These changes take place against the backdrop of the weakening Indian monsoon. The amount of rainfall India now receives is about 6 per cent less than it did in the 1950s. But this reduced rain is now shoehorned into a smaller window of time. Rain is delivered in increasingly intense bursts that lead to flooding, followed by long dry periods with low rainfall. A recent analysis by Prof. Roxy Koll of the Indian Institute of Tropical Meteorology and his colleagues shows that extreme rainfall events have tripled in central India between 1950 and 2015. Alongside, the Indo-Gangetic Plain has become increasingly prone to droughts. During the twenty-first century, the monsoon is expected to strengthen, increasing the rainfall India receives through the year, especially during the south-west advancing monsoon period. Alongside, the risk of floods will increase.

But a worse future lies in wait for India. As climate change progresses, and global warming melts the glaciers of the Himalayas, the melted ice waters will pour into the Ganga, Brahmaputra and Indus. The floods of Pakistan in 2022 are attributed partially to the beginning of the glacier melt that is already underway. More water in the rivers may seem to be a good thing. But in a few decades, once the glaciers have melted and disappeared, these once-massive rivers may become a pale shadow of their once-glorious size, leading to devastating droughts across the Himalayan belt and the Gangetic plains.

The growing urban belt of cities and towns that line the Himalayas will be hard-hit by these changes. Groundwater in the once-fertile Gangetic plains of Punjab, Haryana and Uttar Pradesh, already scarce and under pressure, will not be able to meet the demands of these cities.

Glaciers and monsoons impact the rivers of India, but the rain clouds originate in the oceans. As the surface of the sea becomes warmer, this heat turbocharges atmospheric convection currents and increases the moisture content in the air, pumping more water into the clouds. While ocean temperature has risen by 0.7 degrees C across most of the globe, the waters of the Indian Ocean have heated up even more. This region is now 1 degree C warmer than it used to be in the mid-twentieth century. This explains why the east coast of India has been especially prone to intense coastal storms, along the Bay of Bengal. But now this trend is spreading to the Arabian Sea, which has also begun to show increased warming with an increase in storms on the west coast.

In the past decade, the sea has risen by about 3 cm along the Indian west coast. This may seem like a trivial increase. But because the slope of the coastline is so gradual, a 3-cm rise in sea levels can push seawaters inland by as much as 17 m. Along the east coast, in areas near Kolkata, sea levels have risen even more sharply, by about 5 cm in the past decade. The cities and villages in this part of India will be even worse affected. Storms further exacerbate coastal flooding. As the ocean warms, the winds pick up some of the excess heat energy from the water. The resultant wave surges and gales of wind push seawater farther into the shore, making large parts of the coastline uninhabitable.

With over 7500 km of coastline containing 486 cities and towns and more than 41 million people, India will be especially hard-hit by sea level rise. Mumbai, Kolkata, Chennai, Surat, Kochi and Visakhapatnam— some of India's largest, and wealthiest cities lie on the coast. These cities lie in the direct path of cyclones that can cause devastating damage to people, buildings and the economy. The 2015 floods of Chennai caused over Rs 22,000 crore worth of damage. The potential damage that sea level rise will wreck on our cities will be far worse.

By the end of this century, sea levels may rise as much as 50 cm, exposing 28.6 million people to coastal flooding and damaging assets worth $4 trillion.

Heavy rainfall leads to flash floods in urban areas, as our experience in Chennai, Bengaluru, Mumbai, Ahmedabad, Kolkata, Kochi and other cities shows. Over the past century, sea levels have increased by 0.77 mm per year in Mumbai. If this continues, iconic sites such as Haji Ali and Marine Drive in Mumbai may go underwater by the end of the century. This should make us pause and think about the concrete and money being poured into the development of beachfront properties, under large urban development projects such as the smart cities mission.

Floods and droughts impact everyone in cities but especially affect the poor. Excessive flooding leads to stagnant water and pollutes the city's water supply, increasing the risk of vector-borne diseases like dengue, and water-borne diseases like diarrhoea that kill millions, especially children, across India. Drought, increasing across many rural regions of India, forces people to migrate in large numbers to cities, which lack the infrastructure to house them. By 2050, India's urban population will double, adding 416 million people. Cities need to gear up to manage this level of future growth, along with the additional impacts of climate change.

As sea levels rise, flooding along coasts will displace the tens of millions of people who live in densely packed coastal cities. Where will they go?

Is this what is in store for urban India—an apocalyptic future with drought, storms, floods, displacement, submergence and hunger?

All is not lost. One area of hope is on the issue of aerosol pollution. Research suggests the air pollution and haze that hover around the Indian skies contribute to the weakening of the Indian monsoon. The aerosol layer contributed by air pollution creates a brown surface that acts like an umbrella, reducing the ability of the sun's rays to heat the air above the land. If we tackle India's air pollution, the South Asian monsoon can be revived—a double-benefit that should spur us to act.

Indian cities need to gear up to address challenges of climate change as well. Each of us can play our part here. Enthusiastic educators can play an important role in spreading awareness about climate change among children and adults. Students can take on the task of monitoring the status of local water bodies. Citizen activists can leverage pressure on the government to protect and restore water bodies. Engineers and innovators can develop new local solutions to help cities adapt to climate change. Community leaders and civic society organizations can help make low-income neighbourhoods resilient to climate change by improving water flow, sanitation and access to clean water.

Media plays an especially critical role in shaping policy and government action. Our own research has demonstrated that the Indian parliament asked few questions on climate change between 1999 and 2019, but media reports played a significant role in raising the interest of parliamentarians.

Each of us, in various contexts, has a role to play. But we need to be better informed about the scale and urgency of the challenges that confront us. Knowledge is power. Even if that knowledge appears scary and overwhelming, the very task of action can remove despondency and generate hope.

And there is always hope.

Early Warnings of Climate Change

In the late nineteenth century, the Swedish physicist and Nobel Laureate Svante August Arrhenius published a paper 'On the Influence of Carbonic Acid in the Air upon the Temperature of the Ground'. He quantified how carbon dioxide (called carbonic acid at the time) contributed to the warming of the earth, or the greenhouse effect, modifying climate. His research was largely dismissed, until an American scientist, Charles David Keeling, measured changes in carbon dioxide levels in the remote volcanic mountain of Mauna Loa in Hawaii. Between 1958 and 1961, in a series of recordings, Keeling demonstrated that there had been a steady increase in the concentration of carbon dioxide since the nineteenth century.

Since Keeling's pioneering research, there has been growing evidence that the world's climate is changing, caused by the excessive burning of fossil fuels like coal and petrol. The Intergovernmental Panel on Climate

Change, a body of the United Nations, produces a series of reports on climate change, authored by hundreds of the world's leading scientists. The latest reports warn that the world is on a very dangerous path towards warming, with frightening impacts that can devastate large swathes of the planet by the end of the century. Yet these reports continue to be met with indifference, scepticism and even outright denial in many parts of the world. All the while, the impacts of climate change continue to increase exponentially.

TWENTY

GUWAHATI: CITY OF BEELS AND LANDSCAPED RIVERFRONTS

In 2016, an unusual three-month contest was held in Guwahati city. The winner of this contest was the Ganges River dolphin, which gathered 24,247 votes, surging past the black softshell turtle and the greater adjutant stork with 18,454 and 17,302 votes, respectively. And thus, Guwahati became the first Indian city to get an official animal mascot.

Despite their privileged status as mascot candidates, all three species are dangerously close to local extinction. The reason? Their close connection to water. Nestled between the great Brahmaputra River that lies on its northern fringe and the foothills of the Shillong hills to the south, the emerald-green jewel city of Guwahati owes its existence and identity to the life-giving water that surrounds it. The Ganges River dolphin is found in the Brahmaputra River, while the black softshell turtle and the greater adjutant stork live in the beels (lake-like wetlands) dotted across the landscape. The water that once nurtured these species is now under threat.

The largest city in North-east India, Guwahati is often called the 'Gateway to the North-east', because of its connectivity to other North-eastern states by road, railway and the local airport. These transportation networks have helped the city grow but at a cost—the gradual disappearance of its water bodies and the near-extinction of its aquatic wildlife. Guwahati has never needed a city mascot as badly as it does today.

In origin myths, Guwahati's connection to water is specially emphasized. According to the *Kalika Purana*, an ancient Hindu text, Lord Brahma, father of the Brahmaputra River, created a constellation of stars at the location where the city lies. That is why Guwahati was called Pragjyotishpur (*prakka* meaning ancient, *jyotish* meaning heavenly body and *pura* meaning city). This perhaps also explains why the Ahom dynasty, who later ruled Guwahati, called the Brahmaputra *Nam-dao-phi* (river of the star god). Not just the mighty Brahmaputra, the ponds in the city are also mentioned in local folklore. The iconic Digholi Pukhuri (*pukhuri* means a long pond in Assamese) is described as Duryodhana's pond: the very

Duryodhana who is the antagonistic Kaurava prince in the Hindu epic Mahabharata. Believed to have been richly supplied with water through its connection to seven deep wells, legend has it that the pond was used for the royal wedding of King Bhagadatta's daughter to Duryodhana. The king later extended the pond by connecting it to the Brahmaputra River, creating a long canal for his son-in-law to swim in.

Myths aside, history records that Digholi Pukhuri later played an important role in the history of Guwahati, serving as a naval base and dockyard for the Ahom dynasty, giving them an essential edge in their battles against the Mughals. Centuries later, when the British took over the city, they failed to appreciate the importance of this water body and its connection to the river. The northern portion of the pond was filled in, severing the connection between Digholi Pukhuri and the Brahmaputra. The High Court and Circuit House were later built on this reclaimed land, reducing the pond to a mere 500 m in length.

Despite shrinking in size, Digholi Pukhuri continued to play an important role for Guwahati. The pond that is said to have hosted Duryodhana became a training ground for local swimmers, some of whom went on to compete at the national level. But, in part because it was severed from the river, disrupting the flow of water, the pond gradually became too polluted for swimming. For human swimming, that is. The pond was still frequented by wildlife, including large flocks of birds. In recent years, even this has become proscribed after a boundary fence was installed. A large flock of ducks that used to be seen in Digholi Pukhuri have lost access to the long pond. It is now famous as a site for civic protests and public campaigns, while the ducks must make do with the nearby ponds of Jor Pukhuri.

The twin ponds of Jor Pukhuri were built by the Ahom king Swargadeo Siva Singha in 1720 for the Ugratara Temple. The two ponds were initially one water body, connected to the Brahmaputra River via the Naojan Canal. A road built in British times divided the pond into two, after which it began to be called Jor Pukhuri (a *jodi* or pair of ponds). A poem written by the prominent Assamese poet and dramatist Atul Chandra Hazarika about a goose that lays a golden egg has a mention of this well-known Jor Pukhuri pond.

Bubur ejoni asil raj hanh, Jor Pukhurit sore
Xeijoni hanhe nitou ekota xonar koni pare.

[Bubu had a goose who used to swim in Jor Phukhuri
She used to lay a golden egg every day.]

Near the Jor Pukhuri ponds are located the Nagkata Pukhuri, adjacent to the Sukreshwar Temple in the heart of the city, getting its name from the fact that snakes were once worshipped here (and occasionally sacrificed as well, or so go the tales). Unfortunately, this once-sacred pond is now filled with sewage and is a sad shadow of its magnificent original self. A fourth pond, the Nakonia Pukhuri (nine-cornered pond), was built alongside the Navagraha Temple. An inscription on the side of the tank states that it was dug in 1753 by Tarun Duwara, a minister of the Ahom king Swargadeo Rajeswar Singha. Now called Sil Pukhuri, the pond is ringed with concrete, and the once-sacred waters that were used to bathe the Navagraha idols are green with contamination.

Though the city, as Pragjyotishpur, is mentioned in ancient Hindu texts of the Ramayana, Mahabharata and Puranas, we know little about when Guwahati was first occupied by humans. But some evidence points to the presence of settlements in a distant past. A rock inscription in the Nilachal Hills, close to the city, is from the fifth century. One of the oldest inscriptions found in the North-east, this demonstrates that Guwahati has a recorded human history that extends considerably further than those of the settlements around it.

A more recent discovery tells us the city may be even older than we thought. In the heart of the city, near the Guwahati Press Club, archaeological remains were discovered accidentally in the 1960s, while digging foundations for the Reserve Bank of India building. Now known as the Ambari archaeological site, excavations conducted over several decades from the 1960s recovered pottery and other evidence of human occupation between the seventh and seventeenth centuries. In a 2008–09 excavation, archaeologists made an exciting discovery of a large, deep pond, with brick steps leading down to the base. They also uncovered evidence of foreign ceramics dating between second century BCE and third century CE, suggesting that the site may date back to the Sunga-Kushana period, making it considerably older than previously thought. Heedless of the forethought of the historic settlers who built a large ancient pond to store water, the current administrators of Guwahati have not paid attention to preserving this important historical location. Overgrown with plants, and submerged in stagnant water during the rains, the site 'resembles a gutter choked with overgrown thicket' as a local newspaper, *Sentinel Assam*, commented in 2017.

Like the site in Ambari, the fortunes of Guwahati city have risen and fallen with time. It was the capital of the Kamarupa kingdom, later passing into the hands of various other rulers, eventually becoming an outpost of the Ahom kingdom. Somewhere along the way, the name changed to Guwahati (*guwa* meaning betel nut trees and *hati* meaning market). Several battles for territory were fought along the banks of the Brahmaputra in medieval times, including at the site of the Silsako Bridge in north Guwahati, and at the base of the Itakhuli Hill, along the river. The Ahom kings valued Guwahati for its strategic importance as an important fortification and naval base. But the continuous battles they fought with their neighbours, including the Mughals and the Burmese, took their toll.

By the time the British took over, the infrastructure of Guwahati was in a shambles and the people in poverty. The city was in such poor condition that the British briefly considered moving the local capital to Tezpur.

The decision of the British to stay transformed Guwahati. In 1901, the city covered 4.5 sq. km, and contained 14,244 residents. Today, the city is about fifty times as large as it used to be, covering an area of 217 sq. km. It also contains seventy times the number of people it used to contain, with a population of over a million people.

The city grew, at the expense of its water systems.

Why is water so important for Guwahati anyway? Blessed with abundant water, surely the city can afford to lose a few water bodies to advance development and growth. This may be what planners and developers thought—at least most of them—but it is this flawed thinking that has led to the precarious environmental condition in which Guwahati now finds itself.

Water permeates Guwahati's landscape, creating a unique geography that differentiates this city from all others. The city is a water–land mosaic comprising the Brahmaputra, multiple subsidiary river systems, beels and interlinked wetlands, marshes, swamps and ponds. Perhaps the greatest influence on the city comes from the massive river at its periphery. Originating in Tibet, the Brahmaputra River system flows through parts of Bhutan, India and Bangladesh. The river traverses a mind-boggling length of 2880 km before ending in the Bay of Bengal. Only 28.67 km of the river—less than 1 per cent of its length—runs through the city of Guwahati. This is also where the Brahmaputra is at its narrowest, just 1.2 km wide at the edge of the city.

But if you were to take a plane from the Lokpriya Gopinath Bordoloi International Airport of Guwahati, and fly over the Brahmaputra, the sheer scale of the river becomes apparent. Looking at the seemingly infinite expanse of water, you might even think you were flying over the ocean. Naturally, being next to a river of such immense size greatly modifies the climate of the city of Guwahati. On the one hand, the river helps keep the city cooler than its surroundings and disperses the pollution from traffic. However, in the winter months, the wind that blows from the river into the city brings fine silt and sand impacting the health of residents. The Brahmaputra is used to transport people and supplies, and keeps the city stocked with fish and vegetables that are grown on the riverbanks.

In addition to the river, Guwahati has five natural drainage basins—the Bharlu, Foreshore and Kalmoni basins, and the Silsako and Deepor beels, which connect to the Brahmaputra. The city is also interspersed with ponds. The Ahom kings and other successive rulers are believed to have constructed as many as twenty-five across the city, clearly seeing a value in these water bodies. But as with many other cities across the world, the British who ruled over Guwahati believed that the ponds were sites of stagnant water that spread diseases. John M'Cosh, a British assistant surgeon, wrote in his book *Topography of Assam* (1837) that the ponds were too deep to be drained, and too large to be cleaned of vegetation. He had a suggestion.

> We are aware that the natives of Cashmere convert the surface of their lakes to a useful purpose by first covering them with mats; afterwards strewing them with a stratum of earth, which becomes a permanent floating soil, and sowing them with the seeds of some congenial vegetable. I am of [the] opinion that if the surface of all the pestilential tanks of Gohatti were so dealt with, and the vegetation kept within proper bounds by frequent cutting, that they might be converted to a beneficial purpose, besides being disarmed of much of their pestilential qualities.

Unfortunately, M'Cosh's plan was not followed. Instead, many of the beels and ponds were polluted, encroached, hemmed in with concrete, filled in and built over. By the late twentieth century, the city began to realize the consequences of such wholesale destruction. As a result, the Guwahati Water Bodies (Preservation and Conservation) Act 2008 was enacted to protect the city's water systems from encroachment and damage, and to develop them for ecotourism.

One of the largest and most biodiverse water bodies accorded protection is the Deepor Beel, a permanent freshwater lake that extends across 4000 ha. *Dipa* is a Sanskrit word for 'elephants', and herds of Asian elephants were visitors to the beel. A staggering 232 species of birds, both migratory and residents, have been recorded here. These include endangered species such as the Baer's pochard, of which the largest flock in India was recorded in Deepor Beel in the 1980s. Parts of the beel and its surroundings have been variously designated as protected areas such as a wildlife sanctuary, a bird sanctuary, a Ramsar site and an Important Bird Area.

In addition to being a biodiversity-rich site, the beel is also important for 800 families that fish in the beel, grow rice at the periphery and collect fodder, thus keeping the city of Guwahati supplied with fish, rice, milk and meat. During the harvest festival of Magh Bihu, festive crowds of as many as 2000 people collect at the pond, diving into its waters to catch fish. Fishing was banned when the pond was declared a Ramsar site. In January 2017, Section 144 in the Code of Criminal Procedure was imposed on Deepor Beel during Magh Bihu. Fishermen nevertheless managed to catch 400 kg of fish in one day along with the support of the Deepor Beel Panchpara Samabai Samiti, a cooperative society of the fishing community, in areas of the beel where the prohibitory order did not extend, keeping their customary festive rituals alive.

Government plans to convert Deepor Beel into a site for ecotourism never really took off. But development and degradation continue to pose a threat. A large garbage dump of the Guwahati Municipal Corporation is located just a kilometre from the beel. Large parts of the beel are now polluted by wastewater and covered with invasive aquatic weeds. It has also shrunk in area due to land use conversion. The number of bird and fish species have declined over the years, including the Baer's pochard population. Seasonal flooding has led to the submergence of many interior islands that the birds once used for nesting. The elephants that used to gambol in the wetland, staying for two to three days, now hardly linger on for a few hours. The National Highway 37 and other construction has also severed the connection between the Deepor Beel and the Brahmaputra. A similar story of degradation and loss extends to other beels in the city such as Barsola, Sarusola and Silsako. Even the Bharalu River, a tributary of the Brahmaputra, and its rivulets, Basistha and Bahini, have now become sewers, choked with the city's garbage.

The impacts of the disappearing beels and network of channels has led to an increase in monsoon flooding. Because of its topographic location in a saucer-shaped depression surrounded by hills, the rainwater that falls in the larger region flows down into the city. The mosaic of water bodies, wetlands and channels protected the city, feeding the excess rainwater into the Brahmaputra—until unplanned urbanization choked, silted and destroyed this network.

Several government plans acknowledge the problems associated with the degradation of water bodies in

Guwahati. But development at the cost of the water bodies continues. Mission Flood Free Guwahati was initiated in 2017 to mitigate the impact of floods but has yet to make a visible impact. Guwahati is one of the cities selected under the Smart Cities Mission, a project aimed at the renewal of urban India. Under this project, 696 acres of interconnected water bodies that include Deepor and Borsola beels, Mora Bharalu and Bharalu rivers and the Brahmaputra riverside have been marked for 'development', via the promotion of ecotourism with water sport, river cruises and festivals at the waterfront. The smart city project comes with grand visions of 'creating our own Rambla from Spain, or the Cheonggyecheon Greenway of Seoul, Korea'. But the tree-lined Rambla and the walkway along the Cheonggyecheon stream are located in a very different urban context, where it is far less likely to find large populations of people like the fishers of Deepor Beel who are dependent on the water bodies for their survival.

Food plazas, floating restaurants, open-air theatres and other tourist-focused expansion will further impact the shrinking biodiversity around the city's water bodies. Meanwhile, the capricious Brahmaputra will become increasingly unpredictable under climate change. Building expensive infrastructure in vulnerable flood-prone parts of the city seems neither economically nor ecologically sound. These cities will not be able to restore their water bodies either if they emphasize tourist revenues over ecology.

It would be sad indeed if the city that was the first in India to select an animal mascot lost all its three mascot candidates—the Ganges River dolphin, the black softshell turtle and the greater adjutant stork—whose existence is so closely linked to the city's water bodies. Yet this is precisely what is likely to happen, unless Guwahati adopts a more comprehensive, inclusive, wiser plan of urban development.

Battles on the Brahmaputra

The Saraighat Bridge across the Brahmaputra acts as a vital connection that links the North-east to the mainland. The old bridge gets its name from the legendary Battle of Saraighat, a naval battle between the Ahoms and the Mughals in 1671. Local versions of history describe the anger of the Ahom general Lachit Borphukan, who boarded a warship only to find that his soldiers and oarsmen, fearing defeat, had begun to flee the battle. Despite being ill himself, he attacked four oarsmen with the blunt edge of his sword, throwing them into the river. This turned the tide. The oarsmen re-boarded the ship after others pleaded with the general, and along with six other war vessels, Lachit Borphukan forced back the Mughals. Today, the mayhem of the battle scene may have been replaced by the mayhem of vehicular traffic, and the bridge itself has been torn down and rebuilt. But the site is a memory of what the city owes to the river.

TWENTY-ONE

HEALING WATERS AND HOLY SPIRITS

Life on earth originated in water. Without water, there would be no forests, birds or butterflies; no people, culture and language; no buildings, commerce or computers. Adivasis in India, Native Americans in North America, aboriginals in Australia and indigenous tribes from the African continent worship water. It is also accorded great significance in all the major religions of the world—Eastern religions such as Hinduism, Buddhism and Sikhism, and the Abrahamic religions that include Christianity, Judaism and Islam. Water worship is so ubiquitous, perhaps one should turn the question around and ask—which culture or civilization did *not* worship water?

The Mesopotamian landscape, between the Tigris and Euphrates rivers, is called the cradle of civilization. Originating around 3100 BCE, some of the world's first, extensive, well-developed urban systems can be found here. Water is fundamental to the origin myths of this riverine landscape. The Enūma Eliš, a Babylonian creation myth found in an ancient library in modern-day Iraq, is recorded on seven clay tablets. The story contained here is believed to date back to at least the second century BCE, perhaps even before. It describes the time before creation, with Apsû, the male god who represented the sweet, freshwaters below the ground, and his consort Tiamat, the goddess of the saltwaters. Together, they gave birth to the world.

In an analogous time period in India, the Rig Veda, one of the oldest Vedic texts in Hinduism, recognizes water to be the main element of life itself. In the older Vedic texts (believed to have been composed between 1500 and 1200 BCE), the Vedic god Varuna is associated with justice and truth, the keeper of the moral code. He is also the god of the night sky. By the later Vedas, Varuna is described as the lord of the celestial waters, and in the even later texts of the Puranas, he is the king of the waters and the salty seas, controlling the clouds and rain, commanding rivers to flow, and helping ships travel with safety on the treacherous waters. The ancient Tamil grammar text Tolkāppiyam, which scholars variously date between the second century BCE and the second century CE, describe Kadalon, the god of sea and rain.

In the Old Testament, water plays an important role, coming into being on the very first day of creation. The Book of Exodus says that when Moses led the people of Israel to freedom from slavery, the waters of the Red Sea parted to allow them to pass safely, and then rose again, drowning the Egyptians who pursued them. Water, in these sacred books, is an instrument of God, protecting the innocent and punishing the evil. Jewish law, which traces its origins back to the Book of Moses, the Torah, pays tribute to the harsh desert landscape of its origins by stating that all flowing water—springs, rivers and the sea—are public property. Only confined bodies of water, such as wells, can be privately owned.

For Zoroastrians, water represents health, and is one of the creations of Ahura Mazda, the Wise Lord. The purity of water itself had to be protected—a secondary purifying agent, it was to be used only after ablutions or cleansing was done using cattle urine. Water is worshipped every month, and Zoroastrian fire temples usually have one sacred well containing holy water.

Islam has its roots in the desert, and water is accorded primacy. The Quran states that every living thing is made of water. Thus, water plays a very important role in ritual cleansing before offering prayers. Islam prioritizes the use of water, first for human health, second for domestic animals and third for agriculture and irrigation. In Islam, the Zamzam Well, close to the Kaaba, has some of the holiest waters in the world, believed to have been provided by Allah in response to the pleas of Ibrahim's wife, Haiar, for water for her thirsty son Ismail. Millions of pilgrims visit this area each year, and drink from the waters of the well.

In Christianity, the sacred relationship with humans and water starts early in life, with baptism performed on very young children by sprinkling holy water, believed to cleanse and drive away spirits. While baptism purified individuals, the deluge or the Flood was brought to cleanse the sins of all mankind.

Rivers, lakes, springs and other sites of flowing water are locations to perform ritual sacrifice, seeking the blessings of the gods (and goddesses, for many sacred beings associated with the water are female). Even the gravest of sins can be purified at the site of a river; or so, at least, the ancient epics and texts of many cultures argue. In many world religions, there is an underground river that separates the world we live in from the realm of the dead, like the Styx River in Greek mythology. The spirits of the dead need to traverse this river, which contains malevolent and friendly spirits and creatures to enter heaven or hell. The epic poem, the Iliad, describes a ritual sacrifice at a spring, under a tree, where the warriors observed the killing of an entire family of sparrows, who refused to leave each other, by a serpent. They interpreted this horrific sight, where innocence and fellowship lost out against brutality and violence, as an omen that they would win their battle against Troy.

Cities in India are a potpourri of religions, communities and cultures, found across all sacred rivers: the Ganga, Yamuna, Cauvery, Brahmaputra, Narmada, Godavari and Indus. In most cities today, the rivers are stagnant, filled with garbage, polluted and overgrown with weeds. Yet even in such conditions, water is sacred. Cremation at the city of Varanasi, on the banks of the Ganga is associated with freedom from the cycle of death and rebirth. Dying with one's feet immersed in the waters of the Ganga is believed by many to ensure salvation. The Muslims of Varanasi also call the Ganga *dariyaye-pak* (the holy river), use the ghats for bathing and refer to its waters as *abe jam-jam* (holy water of Mecca–Medina).

But it is not just the large rivers and ever-flowing springs that are worshipped. Smaller, seasonal, enclosed water bodies in cities, including ponds and lakes, are also worshipped in cities across India. The *karaga* festival of Bengaluru is one such unique festival. Held over nine days in March and April, the horticultural community of the Vannhikulyas worship Draupadi, wife of the Pandavas, at this festival. The karaga, an earthen water pot, holds water in which the spirit of Draupadi is invoked. Beginning at the Uppuneerina Kunte, a tiny saltwater pond within the city's Cubbon Park, and later moving to a small tank (a remnant of the once-massive Sampangi Lake), this festival once involved worship at many different water bodies in the city. Even though many of these have now disappeared, the karaga continues to play a very significant cultural role in Bengaluru.

Within the Golden Temple of Amritsar lies the sacred pond (*amrit-sar*, or nectar-filled pond) after which the city was named. There are many stories told of the origin of this tank. In one version, the king of Patti town married his daughter Bibi Rajni to a leper in a fit of anger, after she argued with him, saying that God was a higher power than the king. One day, she brought her husband to the side of the tank and left him there for a while. He saw a pair of fighting crows fall into the water, emerging white in colour. Suspecting that the pond had healing properties, he crawled into it, and was miraculously cured.

When he emerged, healthy and whole, he had another problem on his hands—his wife, who had returned by then, refused to believe that this man, who looked so different, was indeed her husband! They went to the Sikh Guru Ram Das, who explained the miracle to them, and reconciled the couple. Historians say that Guru Ram Das built the lake at the site it is today. Today, it is supplied with freshwater (along with the other lakes of the city) from an underground channel linked to the Upper Bari Doab Canal. Sikhs consider its waters sacred, and whoever touches its water is believed to be blessed with good fortune.

Without the rains, these water bodies would be dry. Agricultural communities track the rain closely, but prayers are also made in Indian cities to the gods of rain, especially in times of drought. A priest, describing the worship of the local rain god Maleraya in his village at the edge of Bengaluru, says:

> The Maleraya puja was performed when there are no rains. We first collect mud from the bed of the lake. It is stamped, shaped into a figurine. This is decorated with leaves of *lakki soppu*. We then carry the idol across the village, and to other nearby villages. I carried it the last time we did the pooja. As we walk, people pour water on the idol, praying for rain. I spin, making the water they pour fall back on them. But the idol made from the mud of the lake never disintegrates. We then hold a feast in our *gunda thope* (wooded grove) for which each of us contribute.

Gangamma, goddess of water, is also worshipped at many lakes in Bengaluru. After an especially good monsoon, when a lake filled with water and overflowed, Gangamma is thanked by floating a boat with a goat and a lamp made of rice flour on the lake or reservoir, while women and men made an an offering of *bagina*, which consists of pulses, rice and other gifts, in a *mora*, a tray-like object made of palm leaves. In 2022, when the waters of the Krishna Raja Sagara Dam overflowed in Mysuru, the Karnataka chief minister made an offering of bagina at the dam, accompanied by his wife, with television cameras and newspaper reporters covering the entire event.

In Uttar Pradesh, Jharkhand and Bihar, Chhath Puja is an important Hindu festival where the Sun God Surya is worshipped along with the goddess of dawn, Usha, locally Chhathi Maiyya. On the first day, the *nahay khay* (bathe and eat), devotees immerse themselves in the nearest water body, taking home some of the water as an offering before the ritual feast. On the third day, during the *sandhya arghya*, or evening offering, they return to the water again, immersing themselves in it as they worship the setting sun. They also light lamps, floating them on the water. Women and men can be seen performing Chhath Puja today even in the toxic frothing waters of the Yamuna near Delhi, a sign that while the river may be physically polluted, it is still considered ritually pure in the eyes of many. Chhath Puja is also performed in the lakes of Bengaluru, where smaller tanks or *kalyani*s (used to immerse idols of Ganesha and Durga at other times of the year) are filled with water for the north Indian migrants to this south Indian city, helping them sustain their traditions of water worship in a new place.

In terms of water worship, nothing can perhaps equal the scale of the Kumbh Mela, held once in twelve years in Prayagraj, Haridwar, Ujjain and Nashik. The 2019 Kumbh Mela in Prayagraj, at the confluence of three holy rivers—the Ganga, Yamuna and (now invisible) Saraswati—set a record for the largest gathering of humans on the planet, with an estimated 140 million people visiting the sacred spot to bathe.

The reverence and sanctity provided to water across Indian religions might lead us to believe that our rivers, lakes and ponds should be protected, conserved and cleaned. Despite the cultural traditions that hold water to be holy, we see sacred rivers like the Ganga, Yamuna and Cauvery polluted with sewage, garbage and industrial effluents, especially in areas close to cities. Despite the many, multi-crore projects to clean up the polluted rivers of India, they continue to have high levels of contamination. Can an appeal to the sacred traditions of water worship act as a starting point to inspire community protection of water bodies?

The Krishnabai Utsav, a riverfront festival held on the banks of the river Krishna in Wai, a town near Pune, offers a story of hope. The festival was first celebrated in the mid-seventeenth century to commemorate the victory of Shivaji, the Maratha ruler, over Afzal Khan, a general in the Adil Shahi dynasty. Held on the riverfront, at its stone *mandapa*s (pillared halls) and ghats, the Krishna River was a crucial backdrop for the festival, which was held to worship the river goddess. As the river became polluted, the physical location of the festival moved away from the riverbank. In recent years, youth groups have spontaneously organized to address this separation of the once-waterfront festival from the river, using the event to organize river-cleaning and conservation activities, seeking to take back the

Krishnabai festival to the riverbanks of the Wai. In Bengaluru, the *kere habba*s (lake festivals), held at the site of many restored lakes, seek to revive old traditions of water worship in a new urban context, making them more secular, with organic food fairs, lake walks and sustainability exhibitions. The festivals seek to bring together the heterogeneous urban communities of long-term residents and migrants who live around the lake to celebrate the importance of water in the city, an attempt at keeping our relationship with water alive, and ever evolving.

Internationally, the right to water was recognized by the United Nations on 28 July 2010, declaring that clean water was essential for human rights. While not legally binding on countries, the recognition is of symbolic value. In 2017, the High Court of Uttarakhand recognized the Ganga and Yamuna rivers as a living person/legal entity, citing Articles 48A and 51 A(g) of the Indian constitution. This ruling was later stayed by the Supreme Court. But many other rivers have been granted the same legal rights as humans. In July 2019, Bangladesh became the first country to give all its rivers the rights accorded to other living entities, via a ruling by the Supreme Court to protect its deltas from further degradation. In 2017, New Zealand granted legal rights to the Whanganui River, following a long battle by the indigenous Māori people, who have fought for this since 1873. In 2021, the Magpie River of Canada was also granted the rights of personhood after sustained campaigns by the indigenous Innu community, for whom the Magpie is the sacred Muteshekau-shipu River. But without legal enforcement and embedding in water law, this growing movement to declare rivers as living beings in the eyes of the law may become ultimately toothless, unless there is a shift in our orientation as a public towards the water.

In India, despite the ubiquitous sight of polluted water in cities, ideas of the sacredness of water have been woven into the fabric of communities. But we are now becoming increasingly distanced from water. No matter if the city's water is polluted. Today, those who can afford it order 'holy water' online for rituals, just like they can buy canisters of air—Himalayan, Alpine, take your pick. We now think of water purely as a commodity we purchase, expecting it to flow 24/7 from distant sources, sometimes hundreds of kilometres away. In doing so, we sever our once-intense spiritual and mental connection to local water bodies such as rivers, tanks and wells.

Like the youth of Wai, the way forward for us is to re-engage with the sacred and the secular aspects of water worship, in a multicultural and diverse way, as once was our approach. Else, it may not be long before the reverence we accord to water is lost, and along with it, the last remnants of any desire to protect water bodies in our cities. We will then have well and truly lost our way.

Of Jalachars and Bhoot-Pret

Even very tiny wells are considered sacred in Indian cities, occupied by gods and spirits called *jalachar* (in contrast to *bhuchar*, the spirits believed to inhabit the earth). In the fascinating book *Folklore on Wells: Being a Study of Water Worship in East and West* (1918), the author R.P. Masani describes a time when wells, which were then a common feature in Mumbai, were closed down across the city by the municipality to reduce the incidence of malaria (because the stagnant water in wells was considered to be a breeding ground for mosquitoes). But many residents objected. They agreed to cover the well with a wire mesh but refused to cover the wells permanently with wood, as this would prevent the rays of the sun (and sometimes the moon) from falling on the water and purifying it. They quoted Parsi, Hindu and Jain religious texts in support of their arguments. Some complained that they fell ill, and in some cases, people even died after their wells were forcibly closed because the spirit in the well objected to being confined. The owner of the famous Edward Theatre (now Edward Cinema) on Kalbadevi Road requested permission from the Malarial Officer to reopen his well because no Indian company was willing to purchase it from him, afraid of the anger of the well spirit. This permission was granted, and the

Malarial Officer, Prof. Roberts, was told that business had improved afterwards.

Some wells were believed to be inhabited by malicious spirits, *bhoot-pret,* which were spirits of those who died near a well, and then lingered to haunt the site, taking regular 'sacrifices'. Thus, says Masani, the malevolent spirit in the well located behind the Bombay Gymkhana was believed to claim at least three suicides every year. But other wells, tanks and rivers were inhabited by benevolent guardian spirits, and a dip in their waters effected miracle cures as has been the belief in cultures across the world.

TWENTY-TWO

POLLUTION AND RESTORATION: THE ETERNAL CYCLE

The presence of rivers is said to have been a critical factor in determining the location of early settlements. In the Fertile Crescent between the banks of the Tigris and Euphrates rivers in Mesopotamia (today's Iraq), some of the first cities originated—including Uruk, the capital city of the hero of epics, Gilgamesh. The cities of the Indus Valley Civilization were built in the fertile land between two rivers—the Indus and the Ghaggar-Hakra.

Yet, some cities like Bengaluru were founded inland, distant from perennial sources of water. Others like Jammu were close to the river but at a higher elevation, making it difficult to access the water that flowed nearby. Cities created ingenious ways to capture water. The *talab*s (ponds) of Jammu were built in the middle of the nineteenth century to capture rainwater for local use. In Mehrauli, one of the older cities of Delhi, Sultan Iltutmish built a tank called the Hauz-e-Sultani or Hauz-e-Iltutmish in the thirteenth century to harvest rainwater. The tank was later repaired by Alauddin Khilji and Feroz Shah Tughlaq. Though the Yamuna was close to Mehrauli, the rugged terrain made it impossible to build canals to convey the water to the city.

In Jaisalmer, the Gadisar Tank, built by Gharsi Rawal in 1367, provided water to the town until 1965. The tank and its catchment area covering 20 sq. km was kept clean, with strict punishments for those found polluting it. Hyderabad's water came from the Hussain Sagar Lake, built in 1562 on a tributary of the Musi River. The *eri*s (tanks) of Chennai, constituted an interconnected system that was linked to the rivers, and the keres of Bengaluru supplied the city with water for its use for centuries before the advent of piped water.

Today, if you go searching for these traditional water bodies, you may not find them, or if you do, you may find that the once-pristine waters that supplied the cities are now polluted with industrial and household wastes, stinking, and overgrown with water weeds. But occasionally, you might stumble across a protected water body in the most unexpected of places. The stepwell of Agrasen ki Baoli is literally a stone's throw away from the crowded bustling Connaught Place in central Delhi—an oasis of calm if you are lucky to be there at a time when no tourists are around. Sitting in a crowded bus in peak traffic, a glimpse of a bird flying high above the water of lakes such as Kaikondrahalli or Agara in Bengaluru, brings a welcome sense of calm to tempers frayed by noise and bright lights. A quiet evening spent on the banks of a temple pond in Chennai, listening to the chime of the temple bells, and inhaling the fragrance of camphor, can melt the stress of a long workday. As residents of cities, we all recognize the importance of urban water bodies, for the health of the city, and to maintain our own strength and sanity.

Restoring water bodies has been on the agenda for years, with river restoration receiving a lot of attention and funds. Many policies and initiatives such as the Water (Prevention and Control of Pollution) Act 1974 and the Ganga Action Plan initiated in 1986 continue to this day. Many crores of rupees have been spent on the cleaning of the Ganga so far. But it is far from complete because it is so challenging. The length of the river, the varied kinds of human pressure on the landscape, and the social and political complexity of the terrain that it flows through make it a very complex issue, one that cannot be solved by technology alone. Many other rivers, including the Yamuna, are in far worse condition. Some, being less prominent, have received less attention.

Water bodies in cities face many kinds of pollution. Irrigated water and rainfall run-off from farms and orchards, rich in fertilizers and pesticides, make their way to urban water bodies. Large cities in India produce close to 40,000 million litres of sewage every day but can treat less than a third of this waste. The excess nitrogen and phosphorus load from fertilizer-rich agricultural run-off and sewage stimulates the growth of water weeds, which give the appearance of a green carpet afloat on many urban water bodies. These weeds, often exotic invasive species, deplete the oxygen in the water, resulting in large-scale fish death. Many lakes also have a high microbial content of pathogenic bacteria. Meanwhile, the soap, detergent and other cleaning agents used by homes make their way into lakes and rivers, generating froth and foam which sometimes catches fire, as famously seen in Bengaluru's Bellandur Lake.

A new type of pollutant is also beginning to make its way into India's urban water bodies—microplastics. These are tiny particles of plastic smaller than 5 mm, which come from the breakdown of larger plastic materials such as bags and masks that are now ubiquitous in cities, as well as directly, from health and beauty products such as many facial cleaners and toothpastes. These very small particles of plastic are so tiny that they can enter the bloodstream when we drink such water, entering our organs where they can trigger inflammation and disease.

Untreated industrial waste, leachate from garbage dumping sites, concentrated drugs from pharmaceutical companies and medical waste from hospitals add to the pollution in our water bodies, making the water toxic with heavy metals and superbugs, and unfit for consumption. But people do consume this toxic water, which also makes its way into the ground, polluting wells and borewells, and ending up in the city's water supply.

Alarmed at the scale of the problem, communities have come together in many cities, working with governments and private trusts to conserve, protect and restore lakes, ponds and streams. Such restoration projects can be seen in Bengaluru, Chennai, Coimbatore, Hyderabad and many other cities. There are many complexities here too. Municipalities tend to seek engineering solutions to the problems of urban water pollution, while communities focus on restoration for urban recreation. These single-minded approaches often lead to outcomes that are neither appropriate for ecology nor for society at large, except for the needs of a privileged minority.

Lakes, ponds, wetlands, rivers, streams and tanks are functioning ecosystems that have co-evolved over centuries in conjunction with human uses such as grazing, fishing and foraging. And a city is a heterogeneous entity, with social, economic and cultural divides. Livelihood to recreation, sacred to social—there are many imaginations of the same water body. Naturalists think of them as sites for bird watching and places to interact with other biodiversity such as spiders, butterflies and insects. A grazer may see a lake as a source of fodder, while women from the area spend their mornings at the side of the lake, foraging for nutritive urban greens. An angling enthusiast may seek a recreational catch of fish, while a fisherman with a contract for the fish in the lake spends his nights at the lake, to catch and prevent the dumping of sewage which can impact his valuable fish. A work-weary resident employed in a software park may want to use the lake nearby to de-stress, while an elderly grandparent wants to amble around the lake, pushing a stroller with a child strapped inside. During festivals such as Durga Puja and Ganesh Chaturthi, even the most polluted of water bodies becomes sacred again, for immersion of the idols.

Lake restoration projects fail to understand this diversity of uses of water, and of cultural imaginations and anticipations. They lay down a fixed set of objectives, and a definite plan of action, seeking a specific end result that is the same in every location. Consultation and collaboration, much-needed values, are often sacrificed because they are considered time consuming and messy. Lake restoration projects also fail to build in the need for flexible course corrections from the approved plan over time. Known in conservation jargon as 'adaptive management', this was an approach that our ancestors who managed these water bodies knew and practised intuitively.

Where and how should a local community interested in protecting a water body begin? Research is key.

Historical documents, maps and government records can be gathered from the archives, and from repositories of different institutions. It is equally important to talk to the communities and residents living in the area. Older residents will know the now-lost channels where water once flowed, the lost wells, ponds and wetlands around the lakes and streams and the different uses and meanings once carried by the water. Conversations with old settlers and recent migrants are needed to understand the many, and varied goals and expectations of restoration. Is the objective to provide drinking water, or serve as a habitat for biodiversity? Will recreation activities be allowed? What about fishing and grazing?

Hydrological assessments will determine how much water can be retained in the peak monsoon, how low the levels can fall to in a prolonged drought and estimate the extent of run-off and inflow in average and extreme years. A map of local land use can identify obstructions that might hinder the flow of water, changing the direction of movement and leading to flooding elsewhere in the landscape. Surveys of channels can highlight locations which need to be desilted, where debris and vegetation need to be removed, or where widening is needed to ensure an unhindered flow of water. Mapping of upstream wetlands, and understanding the soil structure, slope and extent of silt is equally critical.

Lakes and ponds, silted up through the years, need to be drained and desilted for restoration. The municipal agencies tasked with this are often staffed by engineers, who desilt the lake into a soup bowl or a U shape, with the deepest part at the centre of the lake. But this is a problematic approach that fails to appreciate the ecology of the lake. Typically, the upstream end of the lake, where water comes in, is shallow, while the downstream end, from where excess water flows out to join the next lake, is deep. A gentle gradient of this kind allows aquatic flora and fauna of different kinds to flourish. The lake can thus be a habitat for birds such as waders that prefer the reed grass at the shallow end and also kites that fly above the deeper open waters. Restored lakes also tend to be cemented at the edges, or bordered by stone pitching that inhibits percolation of water into the ground instead of the traditional approach of lining the sides with mud and grass.

In Indian cities, many of which lack underground sewage networks, wastewater flows directly into lakes and rivers. Municipal plans for treatment usually involve the construction of sewage treatment plants (STP), which treat polluted water. These STPs require large areas of land, and guzzle electricity. They are often poorly maintained because of the expense involved, or unable to handle the load of sewage. Wetlands play a critical role in the ecological purification of water. Yet they are often treated as wastelands and converted into built spaces. Ironically, in many polluted lakes, the wetlands are filled in and converted into STPs, that soon lie defunct. There are also species of aquatic plants that can be added to water bodies to clean up pollution from sewage. It is much more difficult to treat industrial waste, medical waste and waste from pharmaceutical industries. However, the widespread practice of disposal of toxic waste into urban rivers and lakes needs a complete, well-enforced ban.

Restored lakes often lose parts of the water surface to areas such as a children's park, a walking path or a garden. While these are necessary spaces in cities starved of open areas for the residents, shrinking the water surface of the lake impacts its capacity to recharge water into the ground, and buffer the land from the summer heat. Boating and restaurants may be attractive for people but if allowed, the crowds and activities, with accompanying noise, bright lights, oil spills and garbage, will pollute the lake once again.

Urban water bodies also do not exist in isolation. They are connected to other water bodies by a network of channels enabling the flow of water, entering through wetlands that filter polluted water and treat it before discharging it through channels to enter downstream lakes and ponds. For example, in the case of Bengaluru, initial restoration efforts failed because of a fragmentary approach, where lakes were restored one at a time. The lakes of Bengaluru are not stand-alone entities—they are networked in series. If a lake in the middle of a series is restored, without cleaning the upstream lakes, it can very quickly become polluted again, wasting the time, expense and labour that went into the restoration project.

Water bodies are not private property. Private builders often construct apartments and layouts in such a way that public access is blocked, making them *de facto* private property. In cities such as Bengaluru, when lakes were once leased to private entities, entry fees were charged in some lakes, which many can ill afford to pay. Some lakes are fenced, with security guards who do not permit entry during the middle of the day, a time when many senior citizens or parents with small children may seek to visit the lake. In other instances, the traditional uses of the lakes, such as fishing, grazing cattle and foraging, and fodder collection are prohibited.

Lakes are public commons. Their restoration should enable uses by diverse groups from all sections of society, with different needs and expectations. Especially those who depend on these water bodies for their livelihoods. There will always be conflicts. But the choice of what to prioritize must be done by understanding the trade-offs for each local context and must be done after careful conversations with diverse groups of users.

Lakes may be difficult to restore, but it can be even more challenging to maintain a restored lake in good condition. This involves the frequent removal of invasive weeds, constant vigilance against dumping and the entry of sewage, regular desilting, and the monitoring of biodiversity and water quality. This cannot be done by the government alone. Local communities need to be continuously and closely engaged, along with local fishermen and other guardians of the lake. If lakes are used as sites for education, in conjunction with local schools, this will also enable children to monitor the condition of the lake, and to build formative connections with nature at an early age. In Bengaluru, many communities involved in lake restoration conduct regular kere habbas, lake festivals where they organize activities for children, organic gardening and composting workshops, nature education, photography, and other activities that help to instil an ethic of urban conservation.

None of the above is easy.

If you choose to embark on restoring a water body in your vicinity, it can be messy, time consuming and draining of energy and resources. Even if you do successfully restore the water body, the battle to maintain it is long and hard, requiring constant vigilance against sewage let in by surrounding buildings, garbage dumped by passers-by, encroachments from builders, and pollution from industrial and medical waste. Money is always short for maintenance, but infrastructure keeps falling apart. People who had once committed to protecting the lake leave because they need to move out of the city. Or, as it happens, they simply burn themselves out, after years of effort trying to put out daily fires.

But it is also a joy, a journey, a cause that brings fulfilment and meaning.

Anyone who embarks on restoring a water body also takes heart from the power of the collective, from the fact that the collective action of urban residents can lead to miracles. No matter how different our imaginations of the water body might be, no matter how many stakeholders are involved, no matter how many challenges are thrown up, many water bodies have been restored, in cities across India, because communities that cared have come together with a common goal in mind. That brings hope for the future.

TWENTY-THREE

LAKSHADWEEP: WATER EVERYWHERE, NOR ANY DROP TO DRINK

Lakshadweep is a breathtakingly beautiful collection of islands located in the Arabian Sea around 500 km from Kochi on the south-west coast of India. Once called Laccadives by the Portuguese, the name conjures up a grand vision of a lakh of islands. In reality, this archipelago contains thirty-six islands.

Scattered like rice grains in the middle of the vast ocean, the lagoon area—the shallow water around the islands separated from the oceans by the reefs—covers 4200 sq. km in area. The islands themselves are much smaller, collectively covering just 32.2 sq. km in area and housed within twelve atolls. It would take twenty of these to cover the same area as Mumbai city does.

Each island is even smaller. Viringili, near the reef of the Minicoy Island, is really only an islet, just 2 ha in size, about the size of two football fields placed end to end. The largest island, Andrott, is not vast by any means, though much larger than Viringili, covering 4.84 sq. km. Of these, eleven islands are inhabited, with a population of 64,473 in 2011. Although the population has now probably crossed 70,000, it is still less than 0.1 per cent of the total population of India. Yet as ecologist Rohan Arthur points out, this population is crammed into the small area of the inhabitable islands and has now exceeded 2000 people per sq. km, thrice the national average of India. The islands are, quite literally, running out of space.

And out of water.

For many in India, Lakshadweep is a familiar name, attractive for its visual beauty, and for tourists who throng to the island to experience the underwater through scuba diving and other water sports. For nature lovers, this archipelago of atolls and coral reefs, with its marine life, fishes in every hue of the rainbow and its gentle giants—hawksbills and green turtles that swim in the waters—holds a special lure.

Water is everywhere, enfolding the tourist and the local resident alike in its warm embrace. But only a scant fraction of this water can directly support human life.

The availability of water has always determined island habitability. In 1846, when Lakshadweep was battered by a cyclonic storm, some of the islands witnessed a massive ingress of seawater that contaminated the entire water supply. Those rendered homeless needed to be resettled. One of the locations that was considered, the Suheli islet, was removed from the list because its water was too brackish to drink. The only visitors to this island, fishers from Kavaratti, satisfied their thirst by drinking coconut water.

The islands of Lakshadweep receive an average rainfall of 1600 mm each year. This may seem adequate, except for one fact. Where does a small island, floating on a base of saltwater, store and retain freshwater that is available year-round for islanders and tourists? Monsoon rains that fall on the islands seep through the porous sand and float on a base of saline water, forming a thin layer of freshwater just below the island's surface. Each island has a lens of groundwater, of varying thickness. The size of the freshwater lens is also influenced by the shape of the islands. Rounder islands have larger lenses when compared to the elongated ones. The shape and size of the islands therefore determine how protected the lenses are from the ingress of seawater The separation between fresh, i.e., non-salty rainwater, and salty or alkaline seawater, is not complete. In many places, such as at the edge of the island, the water is brackish or salty. When it rains, the water quality improves, but with the progress of the dry season much of the freshwater tends to evaporate.

Islanders traditionally accessed this water using shallow wells, and every house typically had one or two such wells, which they used throughout the year.

An idyllic description of the island of Minicoy, by the British officer R.H. Ellis in 1924, describes a landscape of wells.

> There is a well in nearly every compound and in the open spaces. The wells are square in shape with plastered sides. Water is found within a short distance of the surface and is drawn by means of a coconut shell tied to a long stick. There are several fine bathing tanks in the village with plastered sides, steps leading down into the water, and parapets. Separate tanks are set apart for the women.

By the late 1990s, there were around 10,000 wells across all islands. This is still a large number considering the population of the islands in 2001 was 60,650—with one well for every six people! But there are no recent estimates that tell us how many of these remain today.

In the islands, with their very small area and narrow width, the locations of the wells determine the type of the water. Wells located close to the coastline tend to see more fluctuations in water quality, especially during the summer months with seawater seeping in.

The presence of the ubiquitous coconut trees also affects water quality. Coconut is an extraordinarily useful tree, central to the life of the islanders. It is one of the few trees that grow well in the sandy soil, windy breeze and challenging climatic conditions of the islands. But while the density of coconut plantations in Lakshadweep is among the highest in the world, the productivity is low. Further coconut trees transpire heavily, demanding a lot of water. In the fragile ecosystems of the Lakshadweep islands, if there are too many coconut trees, the thickness of the lens of freshwater reduces, bringing down the availability of freshwater.

The coral reefs, which act as a buffer between the island and its lagoons and the seawater that surrounds it, are also being eroded. Climate change has led to the warming of the oceans and acidification of the seawater, leading to large-scale coral bleaching and mass death events. Recent cyclones and storms, also increasing in frequency and intensity with climate change, have led to breaches in the coral reefs, leading to the intrusion of saltwater.

Scarcity of drinking water has become such a concern that the local residents have appealed to the National Green Tribunal, which asked the administration to take immediate measures to preserve and manage water resources.

Arthur and his colleagues from Nature Conservation Foundation, who have worked in the islands for over two decades, document an alarming situation where conflicts over freshwater have begun to increase.

Administrators began to address these deficiencies by creating desalination plants on Kavaratti and Minicoy, now seeking to expand to other islands. But these massive plants take up scarce island real estate, and supply less than 10 per cent of the water requirements. With an increased push towards tourism, and the growing challenges of urbanization, infrastructure expansion and the commercial exploitation of fish, which increase the consumption of water, this does not seem sustainable in the least.

No one seems to be viewing the island ecosystem in its entirety—as a system composed of the island, reef and lagoon, along with the thin, fragile, vulnerable lens of groundwater that needs to be conserved at all costs. Increasing waste from tourism and urbanization, pollution, felling of trees and construction of beachfront tourist villas on fragile beach ecosystems have further impacted the already fragile island ecosystems.

Adding to this vulnerability are new proposals to develop Kavaratti, the capital of Lakshadweep, into a smart city to leverage the island's potential as a tourist centre. Part of the Smart Cities Mission that aims to develop 100 cities into smart cities across the country, the project primarily aims to upscale the infrastructure and services of existing cities, with a grand stated vision of emulating cities such as Singapore.

The idea of a smart city is not by any means recent, nor is it Indian in origin. First used in the 1990s, varying definitions of smart cities have been proposed, associating the word 'smart' with meanings such as creative, intelligent, innovative, competitive, sustainable, inclusive, interconnected or efficient. Smart city documents emphasize the potential of technology to address urban challenges, from issues of environment to employment. What does the smart city plan envisage for the fragile, water-scarce island of Kavaratti?

Kavaratti is 5.6 km long, extending across 1.2 km at its widest point. About 11,000 people live on the island. Its pristine sandy beaches and calm lagoons make it an ideal spot for swimming, sunbathing and water sports, making it a major attraction for tourists. The island also has lesser-known but beautiful architectural buildings like the Ujra Mosque, whose interiors are decorated with intricately carved driftwood. The waters of the well in this mosque are believed to have curative properties.

There are about 190 ponds and 1325 open wells in Kavaratti. Private homes have created close to 500 rainwater harvesting structures. But groundwater scarcity and intrusion of seawater have rendered much of the water undrinkable without treatment. In 2005, a desalination plant that was set up on the island brought temporary relief. But water is an ever-present challenge. It is very peculiar then, that the SWOT (Strength, Weakness, Opportunity, Threat) analysis for the Kavaratti smart city project does not mention the challenges of water availability or quality—despite recognizing the ecological fragility of the island, and its vulnerability to natural disasters. The plan does describe water as a focus area. The Sudha Jalam project states that two desalination plants will be set up to supply 2.5 lakh litres per day, and rainwater harvesting will be expanded to collect an additional 10 lakh litres per year.

But these plans need to be seen in the context of the future vision of the island. A vision in which intensive construction is planned—to expand the jetty, build a new community hall and multi-speciality hospital, improve internet services and road infrastructure, and many other projects. In line with the focus of many smart city projects on beautification, the ecosystems of the Kavaratti beaches will be converted through mainly 'innovative' repurposing into a coral museum, botanical garden and waterfront villas for tourists. If/when implemented, these activities will further increase the pressure on the island's scarce water resources.

Keeping in mind the extreme fragility of the islands, and the future threats from climate change, should the focus of the smart city be on investing in the water-security of the island, or on infrastructure and tourist potential?

The twenty-first century will be a century of sea level rise. The islands of Lakshadweep are likely to go underwater by the end of the century, rendering its population perhaps one of India's first internally displaced climate refugees on a large scale. The Lakshadweep Action Plan on Climate Change 2012 states that the islands must 'rationalize the use and management of freshwater resources and adopt improved water conservation practices'. But this expected rationality is not in line with the smart city vision.

Water, water, every where, And all the boards did shrink; Water, water, every where, Nor any drop to drink.

'The Rime of the Ancient Mariner' by the poet Samuel Taylor Coleridge captures the desperation of a sailor stranded at sea, surrounded by saltwater which he cannot drink to quench his unbearable thirst.

This could be the fate of Lakshadweep in a few decades unless an integrated ecological vision can be developed for the island. Such a vision needs to be developed in consultation with its people, and in sync with its traditional livelihoods—instead of being imposed by outside imaginations of economic growth, impelled by tourism.

TWENTY-FOUR

CONNECTING
THE DROPS

Each one of us has experienced thirst—the parched lips, dry mouth and that intense need to drink something. We also know the feeling of relief—how it immediately refreshes the body and mind, almost like magic—when we gulp fresh, cool water and satiate our thirst.

Water is life.

Water is around us, clear and visible in the cities and towns we live in. And yet, we seem to pay it little attention. Absorbed in the business of urban living, we think of water only when there is either too much of it in the city after the rains and floods, or too little in times of drought. In doing so, we forget the many shades of blue that inhabit our cities and bring them alive. Lakes, ponds, streams, rivers and wells do not exist for the sole purpose of supplying us with water for our homes, industries and swimming pools, even though it may seem so at times. Our biodiverse water bodies are home to a variety of fascinating species, with whom we share our planet. Water is integral to Indian cultures, sacred rituals, musical traditions, myths and fables, livelihoods and economies, transport and tourism that constitute the foundation of our daily lives, and the historical underpinnings of our development and growth as a society.

Through the vignettes in this book, we hope to take you on a journey across India, visiting various cities, and discussing a range of themes that showcase the importance of water in our everyday, frenzied, parched lives. We hope that reading this book will pique your interest in the water bodies around you, whether lakes, rivers, streams, tanks, wells, ponds, bheris or wetlands.

We also hope that the stories of inspiration in the book will urge many readers to begin their own journeys of water protection, in whatever form or manner that comes most easily to them. Some of these journeys may begin with introspection on how to conserve water, inspiring action to reduce water consumption. As many water warriors describe, such individual actions often lead to personal transformations, which then prompt people to take on larger collective challenges such as neighbourhood trash clean-ups and lake restoration.

As we take such an approach, we may find that it requires us to shift the way we look at our water bodies and their surroundings, from public goods that the government needs to provide, to urban commons—accessible to all, with each of us bearing a shared responsibility for its care. Finally, there are more complex issues of antimicrobial resistance, tackling industrial pollution or taking a position on large dams. This requires more knowledge—and also that we look at the issue from multiple angles, seeking to understand the consequences of these decisions in our own near neighbourhoods, on distant communities impacted by our decisions and even on future generations.

We seek to provide readers with a palimpsest of stories on water—overlapping, interconnected narratives that showcase the various vibrant facets of water in Indian cities. Taken together, the chapters in the book paint a picture of the different shades of blue that we see around us. Of the range of histories, meanings, memories and imaginations that people ascribe to urban water.

We hope that this book provides a series of samplers that you can either choose to dip into for a refresher, or read at a stretch to quench your thirst, continuing a lifelong journey of engagement with water, the elixir of life.

LIST OF SPECIES

Asian elephant: *Elephas maximus*
Baer's pochard: *Aythya baeri*
Baya weaver bird: *Ploceus philippinus*
Black softshell turtle: *Nilssonia nigricans*
Blue whale: *Balaenoptera musculus*
Bombay duck: *Harpadon nehereus*
Bottlenose dolphin: *Tursiops* spp.
Catla: *Labeo catla*
Checkered keelback: *Fowlea piscator*
Coconut: *Cocos nucifera*
Colossal squid: *Mesonychoteuthis hamiltoni*
Common bottlenose dolphins: *Tursiops truncatus*
Cormorant: *Phalacrocorax* spp.
Duck-billed platypus: *Ornithorhynchus anatinus*
False trevally: *Lactarius lactarius*
Ganges River dolphin: *Platanista gangetica*
Giant squid: *Architeuthis dux*
Giant water bug: *Lethocerus maximus*
Goldspotted grenadier anchovy: *Coilia dussumieri*
Greater adjutant stork: *Leptoptilos dubius*
Greater flamingo: *Phoenicopterus roseus*

Green turtle: *Chelonia mydas*
Hawksbill turtle: *Eretmochelys imbricata*
Hilsa: *Tenualosa ilisha*
Indian black turtle: *Melanochelys trijuga*
Indian flapshell turtle: *Lissemys punctata*
Indian mackerel: *Rastrelliger kanagurta*
Indian oil sardine: *Sardinella longiceps*
Indian softshell: *Nilssonia gangetica*
Koel: *Eudynamys scolopaceus*
Lakki soppu: Vitex negundo
Lesser flamingos: *Phoeniconaias minor*
Marsh crocodile: *Crocodylus palustris*
Mugger: *Crocodylus palustris*
Nilgiris barb: *Hypselobarbus dubius*
Okapi: *Okapia johnstoni*
Olive Ridley turtle: *Lepidochelys olivacea*
Peacock: *Pavo cristatus*
Peregrine falcon: *Falco peregrinus*
Saltwater crocodile: *Crocodylus porosus*
Tigertooth croaker: *Otolithes ruber*
Whistling thrush: *Myophonus* spp.
Yangtze giant softshell turtle: *Rafetus swinhoei*
Yangtze River dolphin: *Lipotes vexillifer*

SELECTED REFERENCES

To keep the reading experience more enjoyable, we have avoided providing an extensive list of citations in the text. For interested readers, we provide a list of major sources below. To keep this list manageable in size, we have excluded newspaper and internet articles, from which we may have taken well-known information or descriptions of events.

1. **Blue Waters**
i.   Igor A. Shiklomanov, 'World Freshwater Resources', in *Water in Crisis: A Guide to the World's Fresh-water Resources* (Oxford University Press, 1993).

2. **Delhi: Re-Imagining the Yamuna**
i.   Alexander Follmann, *Governing Riverscapes: Urban Environmental Change along the River Yamuna in Delhi, India* (Franz Steiner Verlag, 2016).
ii. Amita Baviskar, 'What the Eye Does Not See: The Yamuna in the Imagination of Delhi', *Economic & Political Weekly*, vol. 46, no. 50 (2021): 45–53.
iii. Anon, *Orders Prohibiting the Grant of Land in the River Bed Opposite the Bela for Melon Cultivation*, File from Department of Delhi Archives, 1917.

iv. Anon, *Correspondence between Deputy Commissioners Office and Delhi Improvement Trust May-June-1945 on Grant of Land for Swimming and Rescue Club to Hindu Scouts Association*, File from Department of Delhi Archives, 1945.

v. Anon, *Rescue Arrangements at Jamuna River*, File from Department of Delhi Archives, 1949.

vi. Awadhendra Sharan, *In the City, Out of Place: Nuisance, Pollution, and Dwelling in Delhi, c. 1850–2000* (Oxford University Press, 2014).

vii. Iñaki Alday and Pankaj Vir Gupta, *Yamuna River Project: New Delhi Urban Ecology* (ACTAR Publishers, 2018).

viii. Pushp Jain, *Sick Yamuna, Sick Delhi: Searching a Correlation* (PEACE Institute Charitable Trust, 2009).

3.　Walking on Water

i. James H. Jhonston, *Precis of Reports, Opinions and Observations on the Navigation of the Rivers of India by Steam Vessels* (J.L. Cox [Printers], 1831).

ii. Arthur Cotton, *Public Works in India: Their Importance with Suggestions for Their Extension and Improvement* (Richardson Brothers, 1854).

iii. Bejoy Kumar Sarkar, *Inland Transport and Communication in Medieval India* (Calcutta University Press, 1925).

iv. D.C. Phillott, *The Ain-I Akbari (by Abu 'L-Fazl Allami)* (The Asiatic Society of Bengal, 1927).

v. James Rennell, *An Account of the Ganges and Burrampooter River* (J. Nicholas [Printers], 1781).

vi. Jenia Mukherjee, *Blue Infrastructures: Natural History, Political Ecology and Urban Development in Kolkata* (Springer, 2020).

vii. John Bourne, *Indian River Navigation: A Report* (W.H. Allen & Co., 1849).

viii. L.N. Rangarajan, *The Arthashastra* (Penguin Books, 1987).

ix. Robert Ivermee, *Hooghly: The Global History of a River* (HarperCollins, 2021).

x. Rowan Hackman, *Ships of the East India Company* (World Ship Society, 2001).

4. Of Drugs and Superbugs

i. Adam M. Schaefer, Gregory D. Bossart, Tyler Harrington, Patricia A. Fair, Peter J. McCarthy and John S. Reif, 'Temporal Changes in Antibiotic Resistance among Bacteria Isolated from Common Bottlenose Dolphins (*Tursiops truncatus*) in the Indian River Lagoon, Florida, 2003–2015', *Aquatic Mammals*, vol. 45, no. 5 (2019): 533–42.

ii. Ajai R. Singh, 'Science, Names: Giving and Names Calling: Change NDM-1 to PCM', *Mens Sana Monographs*, vol. 9, no. 1 (2011): 294–319.

iii. Alberto Giubilini, 'Antibiotic Resistance as a Tragedy of the Commons: An Ethical Argument for a Tax on Antibiotic Use in Humans', *Bioethics, Special Issue on Ethics of Antibiotic Resistance*, vol. 33 (2019): 776–84.

iv. Babu Rajendran Ramaswamy, Govindaraj Shanmugama, Geetha Velua, Bhuvaneshwari Rengarajana and D.G. Joakim Larsson, 'GC–MS Analysis and Ecotoxicological Risk Assessment of Triclosan, Carbamazepine and Parabens in Indian Rivers', *Journal of Hazardous Materials*, vol. 186, no. 2/3 (2011): 1586–93.

v. D.G. Joakim Larsson, Cecilia de Pedro and Nicklas Paxeus, 'Effluent from Drug Manufactures Contains Extremely High Levels of Pharmaceuticals', *Journal of Hazardous Materials*, vol. 148, no. 3 (2007): 751–55.

vi. Dongeun Yong, Mark A. Toleman, Christian G. Giske, Hyun S. Cho, Kristina Sundman, Kyungwon Lee and Timothy R. Walsh. 'Characterization of a New

Metallo-Beta-Lactamase Gene, Blandm-1, and a Novel Erythromycin Esterase Gene Carried on a Unique Genetic Structure in *Klebsiella pneumoniae* Sequence Type 14 from India', *Antimicrobial Agents and Chemotherapy*, vol. 53, no. 12 (2009): 5046–54.

vii.   Manisha Lamba, T.R. Sreekrishnan and Shaikh Ziauddin Ahammad, 'Sewage Mediated Transfer of Antibiotic Resistance to River Yamuna in Delhi, India', *Journal of Environmental Chemical Engineering*, vol. 8, no. 1 (2018): 102088 (Online).

viii. Neelam Taneja and Megha Sharma, 'Antimicrobial Resistance in the Environment: The Indian Scenario', *Indian Journal of Medical Research*, vol. 149, no. 2 (2019): 119–28.

ix.   Onkar A. Naik, Ravindranath Shashidhar, Devashish Rath, Jayant R. Bandekar and Archana Rath, 'Characterization of Multiple Antibiotic Resistance of Culturable Microorganisms and Metagenomic Analysis of Total Microbial Diversity of Marine Fish Sold in Retail Shops in Mumbai, India', *Environmental Science and Pollution Research*, vol. 25, no. 7 (2017): 6228–39.

x.   Safaa Altves, Hatice Kübra Yildiz and Hasibe Cingilli Vural, 'Interaction of the Microbiota with the Human Body in Health and Diseases', *Bioscience of Microbiota Food Health*, vol. 39, no. 2 (2019): 23–32.

xi.   Society for Healthcare Epidemiology of America, Infectious Diseases Society of America and Pediatric Infectious Diseases Society, 'Policy Statement on Antimicrobial Stewardship by the Society for Healthcare Epidemiology of America (SHEA), the Infectious Diseases Society of America (IDSA), and the Pediatric Infectious Diseases Society (PIDS)', *Infection Control and Hospital Epidemiology*, vol. 33, no. 4 (2012): 322–27.

xii. Sumanth Gandra, Jyoti Joshi, Anna Trett, Anjana Sankhil Lamkang and Ramanan Laxminarayan, *Scoping Report on Antimicrobial Resistance in India* (Center for Disease Dynamics, Economics & Policy, 2017).

xiii. William Hall, Anthony McDonnell and Jim O'Neill, *Superbugs: An Arms Race against Bacteria* (Harvard University Press, 2018).

5. Mumbai: Wresting Land from the Ocean

i. A.D. Pulsalker and V.G. Dighe, *Bombay: Story of the Island City* (All India Oriental Conference, 1949).

ii. Anon, *Report of the Committee Appointed by the Government of India to Enquire into the Bombay Back Bay Reclamation Scheme* (His Majesty's Stationery Office, 1926).

iii. George Curtis, 'The Development of Bombay', *Journal of the Royal Society of Arts*, vol. 69, no. 3582 (1921): 559–77.

iv. Govind Narayan, *Govind Narayan's Mumbai: An Urban Biography from 1863* (Anthem Press, 1863).

v. Hemantkumar A. Chouhan, D. Parthasarathy and Sarmishtha Pattanaik, 'Coastal Environmental Vulnerability: Sustainability and Fisher Livelihoods in Mumbai, India', in *Towards Coastal Resilience and Sustainability* (Routledge, 2019).

vi. Hemantkumar A. Chouhan, D. Parthasarathy and Sarmistha Pattanaik, 'Urban Development, Environmental Vulnerability and CRZ Violations in India: Impacts on Fishing Communities and Sustainability Implications in Mumbai Coast', *Environment, Development and Sustainability*, vol. 19 (2017): 971–85.

vii. J. Gerson Da Cunha, *The Origin of Bombay* (Asian Educational Services, 1900).

viii. Mariam Dossal, *Imperial Design and Indian Realities: The Planning of Bombay City, 1845–1875* (Oxford University Press, 1991).

ix. R.X. Murphy, 'Remarks on the History of Some of the Oldest Races Now Settled in Bombay; With Reasons for Supposing that the Present Island of Bombay Consisted in the 14th Century of Two or More Distinct Islands', *Transactions of the Bombay Geographical Society 1836–38,* (1838): 128–39.

x. Samuel T. Sheppard, *Bombay* (The Times of India Press, 1932).

xi. Shibaji Bose, Upasona Ghosh, Hemant Kumar Chauhan, N.C. Narayanan and D. Parthasarathy, 'Uncertainties and Vulnerabilities among the Koli Fishermen in Mumbai', *Indian Anthropologist*, vol. 48, no. 2 (2018): 65–80.

xii. Tim Riding, '"Making Bombay Island": Land Reclamation and Geographical Conceptions of Bombay, 1661–1728', *Journal of Historical Geography*, vol. 59 (2018): 27–39.

6. Fantastic Beasts and Water Monsters

i. Brian Regal, *Searching for Sasquatch: Crackpots, Eggheads, and Cryptozoology* (Palgrave Macmillan, 2011).

ii. Conrad Bauer, *Paranormal Creatures: Investigating Cryptozoology* (Createspace Independent Publishing Platform, 2015).

iii. Darren Naish, *Hunting Monsters: Cryptozoology and the Reality behind the Myths* (Arcuturus Publishing Limited, 2016).

iv. Deena West Budd, *The Weiser Field Guide to Cryptozoology* (Red Wheel/WeiserBooks, 2010).

v. George M. Eberhart, *Mysterious Creatures: A Guide to Cryptozoology* (ABC-CLIO, 2002).

vi.   George M. Williams, *Handbook of Hindu Mythology* (ABC-CLIO, 2003).

vii. Loren Coleman and Jerome Clark, *Cryptozoology A to Z: The Encyclopaedia of Loch Monsters, Sasquatch Chupacabras and Other Authentic Mysteries of Nature* (Simon & Schuster, 1999).

viii. Lorenzo Rossi, 'A Review of Cryptozoology: Towards a Scientific Approach to the Study of "Hidden Animals', in *Problematic Wildlife* (Springer International Publishing, 2015).

ix.   Peter Scott and Robert Rines, 'Naming the Loch Ness Monster', *Nature,* 258 (1975): 466–68.

7.   Is All Well in Our Cities?

i.   Anil Agarwal and Sunita Narain, *Dying Wisdom: Rise, Fall and Potential of India's Traditional Water-Harvesting Systems* (Centre for Science and Environment, 1997).

ii.   Ehud Galili and Yaacov Nir, 'The Submerged Pre-pottery Neolithic Water Well of Atlit-Yam, Northern Israel, and Its Paleoenvironmental Implications', *Holocene,* vol. 3, no. 3 (1993): 265–70.

iii. Hita Unnikrishnan and Harini Nagendra, 'From Pulley to Pipe: The Decline of the Wells of Bangalore', *Arcadia,* vol. 5 (2018) (Online).

iv.   L.N. Rangarajan, *The Arthashastra* (Penguin Books, 1987).

v.   Michal Rybníček, Petr Kočár, Bernhard Muigg, Jaroslav Peška, Radko Sedláček, Willy Tegel and Tomáš Kolář, 'World's Oldest Dendrochronologically Dated Archaeological Wood Construction', *Journal of Archaeological Science,* vol. 115 (2020): 105082 (Online).

vi.   Purnima Mehta Bhatt, *Her Space, Her Story* (Zubaan, 2014).

vii. R.P. Taylor, 'Reduction of Physiological Stress Using Fractal Art and Architecture', *Leonardo*, vol. 39, no. 3 (2006): 245–51.

viii. Sandhya Iyengar, *Waternama: A Collection of Traditional Practices for Water Conservation and Management in Karnataka* (Communication for Development and Learning, 2007).

ix. Willy Tegel, Rengert Elburg, Dietrich Hakelberg, Harald Stäuble and Ulf Büntgen, 'Early Neolithic Water Wells Reveal the World's Oldest Wood Architecture', *PLOS ONE*, vol. 7, no. 12 (2012): e51374 (Online).

8. Kolkata: Transforming Waste to Wealth

i. Alexander Hamilton, *A New Account of the East India Being the Observations and Remarks of Alexander Hamilton* (Volume 2) (John Molmon, One of His Majesty's Printers, 1744).

ii. Anon, 'The Sanitary Aspects of the New System of Drainage of Calcutta', *Lancet*, vol. 100, no. 2560 (1872): 415–16.

iii. Anon, *Annual Report on the Administration of the Bengal Presidency for 1867–68* (Bengal Secretariat Press, 1868).

iv. Anon, *Annual Report on the Administration of the Bengal Presidency for 1868–69* (Bengal Secretariat Press, 1869).

v. Assa Doron and Robin Jeffery, *Waste of a Nation: Garbage and Growth in India* (Harvard University Press, 2018).

vi. Calcutta Metropolitan Planning Organisation, *First Report* (Calcutta Metropolitan Planning Organization, 1962).

vii. Dhrubajyoti Ghosh and Susmita Sen, 'Ecological History of Calcutta's Wetland Conversion', *Environmental Conservation*, vol. 14, no. 3 (1987): 219–26.

viii. David Smith, *Report on the Drainage and Conservancy of Calcutta* (Bengal Secretarial Press, 1869).

ix. Development and Planning (Town and Country Planning) Department, *Memorandum on Development Plan Calcutta Metropolitan District 1966–71* (Government of West Bengal, 1965).

x. F.P. Strong, *Excerpts from the Topography and Vital Statistics of Calcutta, Embracing Observations on these Subjects Formed at Different Periods, and Officially Submitted to the Governor General Lord William Bentinck and the Local Authorities by F.P. Strong, Surgeon, 24 Pergunnahs, dated 1828, 1837* (Publisher not known, 1838).

xi. Fever Committee Report, *Appendix F to Report of the Committee upon the Fever Hospital Municipal Improvements Containing Miscellaneous Evidence and Papers* (Bishops College Press, 1839).

xii. Fever Committee Report, 'Captain Prinsep's Memorandum on the Salt-water Lakes, in the Vicinity of Calcutta: With Suggestions for Filling Them up by Warping, Territorial Department, Revenue Consultation, dated 16th February 1830', in *Appendix G to Report of Committee upon the Fever Hospital and Municipal Improvements Containing a Copy of the Minute, on the Drainage of the Salt Water Lake, by the Governor General* (Lord William Cavendish Bentinck, KCB) in the Financial and Revenue Department dated 2nd February 1830 and its connected papers (Bishops College Press, 1841).

xiii. Fever Committee Report, 'Appendix to Foregoing Memoir, of J. Augs Schalch, Lieut, Deputy Assistant Quarter Master General Dated June 1821, Calcutta', in *Appendix to Appendix G of Report of Committee upon the Fever Hospital and Municipal Improvements*

Containing Miscellaneous Evidence and Papers (Bishop College Press, 1841).

xiv. Fever Committee Report, 'Second Appendix to Foregoing Memoir Submitted by J.A. Schalch, Lieut, Deputy Assistant Quarter Master General Submitted by Him to the Committee Appointed to Report Upon His Plan dated 23rd September 1821', in *Appendix to Appendix G of Report of Committee upon the Fever Hospital and Municipal Improvements containing Miscellaneous Evidence and Papers* (Bishop College Press, 1841).

xv. Government of West Bengal, *Reclamation of Land for Building Purposes Near Calcutta.* (Government of West Bengal, 1958).

xvi. H. Chattopadhyaya, *From Marsh to Township East of Calcutta: A Tale of Salt Water Lake and Salt Lake Township (Bidhan Nagar)* (K.P. Bagchi and Company, 1990).

xvii. J. Long, *Unpublished Record of Government for the Years 1847 to 1767 Inclusive Relating Mainly to the Social Condition of Bengal: With a Map of Calcutta in 1784, Vol. 1* (Office of the Superintendent of Government Printing, 1869).

xviii. J.R. Martin, *Official Report of the Medical Topography and Climate of Calcutta with Brief Note of Its Prevalent Diseases Both Endemic and Epidemic* (G.H. Huttmann, Bengal Military Orphan Press, 1839).

xix. Jenia Mukherjee, *Blue Infrastructures: Natural History, Political Ecology and Urban Development in Kolkata* (Springer, 2020).

xx. Local Self Government Department, *Report of the Drainage Outfall Committee (Appointed by the Government of Bengal in the Local Self-Government Department under Resolution 1732 P.H., Dated 29 July 1931)* (Bengal Secretariat Book Depot, 1933).

xxi. Ranjit Sen, *Birth of a Colonial City* (Routledge India, 2019).

xxii. W. Clark, *Report to the Municipal Commissioners on a System of Sewers for the Drainage of Calcutta* (Sanders, Cones and Co., 1855).

9. Tinker, Tailor, Mapper, Spy: Secret Expeditions to Map Rivers

i. Anon, 'Appendices A: Extract from Captain T.G. Montgomerie's letter to the Secretary of the Bengal Asiatic Society', *Journal of the Royal Geographical Society*, vol. 36 (1866): 166–67.

ii. Derek Waller, *The Pundits: British Exploration of Tibet and Central Asia* (The University Press of Kentucky, 1990).

iii. H. Trotter, 'Account of the Pundit's Journey in Great Tibet from Leh in Ladakh to Lhasa, and of His Return to India via Assam', *Journal of the Royal Geographic Society*, vol. 47 (1877): 86–36.

iv. Harini Nagendra, 'Mapmakers 3. Techniques of Cartography', *Resonance*, vol. 4 (1999): 8–15.

v. Indra Singh Rawat, *Indian Explorers of the 19th Century* (Ministry of Information and Broadcasting, Government of India, 1983).

vi. J.A. Field, 'The History of the Exploration of the Upper Dihong', *Geographical Journal*, vol. 41, no. 3 (1913): 291–93.

vii. James Rennell, *Memoir of a Map of Hindoostan or the Mogul Empire* (M. Brown, 1788).

viii. Kenneth Mason, 'Kishen Singh and the Indian Explorers', *Geographical Journal*, vol. 62, no. 6 (1923): 429–40.

ix. L.A. Waddell, 'The Falls of the Tsang-po (San-pu), and Identity of That River with the Brahmaputra', *Geographical Journal*, vol. 5, no. 3 (1895): 258–59.

x. Matthew H. Edney, *Mapping an Empire: The Geographical Construction of British India, 1765–1843* (The University of Chicago Press, 1999).

xi. Oyndrila Sarkar, 'Science, Surveying and Scientific Authority: The Brothers Schlagintweit in "India and High Asia", 1854–1857', *South Asia: Journal of South Asian Studies*, vol. 40, no. 3 (2017): 544–65.

xii. Riaz Dean, *Mapping the Great Game: Explorers, Spies and Maps in 19th-Century Asia* (Casemate Publishers, 2019).

xiii. Sekhar Pathak and Uma Bhatt, *Asia ke Peeth Par: Pundit Nain Singh Rawat* (Pahar, 2007).

xiv. T.G. Montgomerie, 'Report of a Route Survey Made by Pundit *____from Nepal to Lhasa and Thence through the Upper Valley of the Brahmaputra to Its Source', *Journal of the Royal Geographic Society*, vol. 38 (1868): 129–219.

xv. T.G. Montgomerie, 'Report of the Trans-Himalayan Explorations during 1867', *Journal of the Royal Geographic Society*, vol. 39 (1869): 147–87.

10. Shackling the Waters: Dams and Cities

i. Arun Kumar Nayak, 'Development, Displacement and Justice in India: Study of Hirakud Dam', *Social Change*, vol. 43, no. 3 (2013): 397–419.

ii. Bradford Morse and Thomas R. Berger, *Sardar Sarovar: Report of the Independent Review* (International Environmental Law Research Centre, 1992).

iii. Central Water Commission Dam Safety Organisation, *National Register of Large Dams* (Government of India, 2019).

iv. Christopher V. Hill, *South Asia: An Environmental History* (ABC-CLIO, 2008).

v.   Ivan B.T. Lima, Fernando M. Ramos, Luis A.W. Bambace and Reinaldo R. Rosa, 'Methane Emissions from Large Dams as Renewable Energy Resources: A Developing Nation Perspective', *Mitigation and Adaptation Strategies for Global Change*, vol. 13 (2007):193–206.

vi.   Parshuram Ray, 'Development Induced Displacement in India', *SARWATCH (South Asian Refugee Watch)*, vol. 2, no.1 (2000): 33–40.

vii. Ramya Swayamprakash, 'Fishing the Cauvery River: How Mettur Changed It All', *SANDRP (South Asia Network on Dams, Rivers and People)*, 7 June 2014.

viii. Silky Agrawal, Mantu Majumder, Ravindra Singh Bish and Amit Prashant, 'Archaeological Studies at Dholavira using GPR', *Current Science*, vol. 114, no. 4 (2018): 879–87.

ix.   World Commission on Dams, *Dams and Development: A New Framework for Decision-Making* (Earthscan Publications Ltd, 2000).

11. Udaipur: City of Lakes

i.   Anon, *Report on the Administration of Mewar State for Years 1940, 1941 and 1942* (Madras Law Journal Press, 1944).

ii.   Captain J.C. Brookes, *History of Mewar* (Baptist Mission Press, 1859).

iii. Deeksha Dave, 'Eutrophication in the Lakes of Udaipur City: A Case Study of Fateh Sagar Lake', International Conference on Biotechnology and Environment Management IPCBEE (IACSIT Press, 2011).

iv.   Dev Nath Purohit, *Mewar History: Guide to Udaipur* (Times of India Press, 1938).

v.   Fateh Lal Mehta, *Handbook of Mewar and Guide to Its Principal Objects of Interest* (The Times of India Steam Press, 1888).

vi. James Fergusson, *Ancient Architecture Hindostan* (Joseph Hogarth, 1848).

vii. K.D. Erskine, *The Mewar Residency* (Scottish Mission Industries. Co. Ltd, 1908).

viii. Narpat Singh Rathore, 'A Historical Perspective of the Development of Rain Water Harvesting Techniques in the Mewar Region, Udaipur, Rajasthan, India', *International Journal of Water Resources and Arid Environments*, vol. 1, no. 4 (2011): 285–94.

ix. Neha Singh and N.C. Narayanan, 'Rural Urban Water Transfers Conflict over the Mansi Wakal Dam in Udaipur, Rajasthan', in *Conflicts Around Domestic Water and Sanitation in India* (Forum for Policy Dialogue on Water Conflicts in India, 2014).

x. Neha Singh, D. Parthasarathy and N.C. Narayanan, 'Contested Urban Waterscape of Udaipur', in *Sustainable Urbanisation in India: Challenges and Opportunities* (Springer, 2018).

xi. V.V.S. Gurunadha Rao and Mogali J. Nandan, *Udaipur Lakes. Encyclopaedia of Lakes and Reservoirs 2012 Edition* (Springer, 2012).

12. Songs of the River

i. Catherine Grant, 'Climate Justice and Cultural Sustainability: The Case of Etëtung (Vanuatu Women's Water Music)', *Asia Pacific Journal of Anthropology*, vol. 20, no. 1 (2019): 42–56.

ii. Esha Shah, 'Seeing like a Subaltern: Historical Ethnography of Pre-modern and Modern Tank Irrigation Technology in Karnataka, India', *Water Alternatives*, vol. 5, no. 2 (2012): 507–38.

iii. Henrik Brumm and Marc Naguib, 'Environmental Acoustics and the Evolution of Bird Song', *Advances in the Study of Behaviour*, vol. 40 (2009): 1–33.

iv. Padma Seshadri and Padma Malini Sundararaghavan, *It Happened Along the Kaveri: A Journey through Space and Time* (Niyogi Books, 2012).

v. R. Krishnamurthy, *The Saints of Cauvery Delta* (Concept Publishing Company, 1979).

vi. Thomas Dick, 'Vanuatu Water Music and the Mwerlap Diaspora: Music, Migration, Tradition and Tourism', *AlterNative: An International Journal of Indigenous Peoples*, vol. 10, no. 4 (2014): 392–407.

vii. William J. Jackson, *Tyagaraja: Life and Lyrics* (Oxford University Press, 1992).

13. Interlinking Rivers

i. Anasuya Syam and Sushma Sosha Philip, 'Investigating Interlinking: A Critique of India's National River Linking Plan', *Environmental Law and Practice Review*, vol. 5 (2017): 1–21.

ii. Kelly D. Alley, 'The Making of a River Linking Plan in India: Suppressed Science and Spheres of Expert Debate', *India Review*, vol. 3, no. 3 (2004): 210–38.

iii. Pallava Bagla, 'India Plans the Grandest of Canal Networks', *Science*, vol. 345, no. 6193 (2014): 128.

iv. Ramaswamy R. Iyer, 'River Linking Project: A Disquieting Judgment', *Economic & Political Weekly*, vol.47, no.14 (2012): 33–40.

v. Richard Stone and Hawk Jia, 'Going against the Flow', *Science*, vol. 313, no. 5790 (2006): 1034–37.

vi. Rohan D'Souza, 'Canal Irrigation and the Conundrum of Flood Protection: The Failure of the Orissa Scheme of 1863 in Eastern India', *Studies in History*, vol. 19, no. 1 (2003): 41–68.

vii. Rohan D'Souza, 'River Linking and Its Discontents: The Final Plunge for Supply Side Hydrology in India', in *Water First: Issues and Challenges for Nations and Communities in South Asia* (Sage Publications, 2008).

viii. Tushaar Shah and Upali A. Amarasinghe, 'River Linking Project: A Solution or Problem to India's Water Woes?' in *Indian Water Policy at the Crossroads: Resources, Technology and Reforms* (Springer, 2016).

14. Chennai: City of Floods and Drought

i. Alexander Hamilton, *A New Account of the East India Being the Observations and Remarks of Alexander Hamilton* (Vol. 1) (Asian Educational Services, 1739).

ii. Anantanarayanan Raman and Natarajan Meenakshisundaram, 'Water-distribution efforts in Madras: From Sailor George Baker (1750s) to Engineers John Jones, Hormusji Nowroji and James Madeley (1870s–1920s)', *Current Science*, vol. 120, no. 3 (2021): 575–85.

iii. Fanny Emily Penny, *Fort St. George* (Publisher not known, 1900).

iv. Government of Tamil Nadu. *Report of the Comptroller and Auditor General of India on Performance Audit of Flood Management and Response in Chennai and Its Suburban Areas for the Year Ended March 2016. Report No. 4 Of 2017* (Government of Tamil Nadu, 2017).

v. Jayshree Vencatesan, 'Protecting Wetlands', *Current Science*, vol. 93, no.10 (2007): 288–90.

vi. Krupa Ge. *Rivers Remember: The Shocking Truth of a Manmade Flood* (Context, 2019).

vii. Pushpa Arabindoo, 'Mobilising for Water: Hydro-politics of Rainwater Harvesting in Chennai', *International Journal of Urban Sustainable Development*, vol. 3, no. 1 (2011): 106–26.

viii. Rao Sahib C.S. Srinivasachari, *History of the City of Madras* (P. Varadachary and Co, 1939).

ix. S. Muthiah, *A Madras Miscellany: A Decade of People, Places and Potpourri* (Westland Ltd, 2011).

x. Tania Sebastian, 'A "Chennai" in Every City of the World: The Lethal Mix of the Water Crisis, Climate Change, and Governance Indifference', *Law, Technology and Humans*, vol. 4, no. 1 (2022): 79–101.

xi. The Nature Conservancy and Care Earth Trust, *Restoring Chennai's Wetlands* (The Nature Conservancy Centre and Care Earth Trust, no date).

15. Bone Swallowers, Corpse-Eating Turtles and Crocodiles in the City

i. Arvind Singh, 'Save the Gangetic Dolphin to Save the Ganges river', *Science India*, (February 2016): 5–11.

ii. Emma Roberts, *Scenes and Characteristics of Hindostan, Vol. I and II* (W.M.H. Allen and Co., 1835).

iii. Gaurav Kailash Sonkar, Kumar Gaurav and Rajiv Sinha, 'Eco-geomorphic Assessment of the Varanasi Turtle Sanctuary and Its Implication for Ganga River Conservation', *Current Science*, vol. 116, no. 12 (2019): 2063–71.

iv. Howard Anderson Musser, *Jungle Tales* (George H. Doran Company, 1922).

v. J.T. Polhemus, 'Hemiptera (True bugs)', in *Encyclopaedia of Inland Waters* (Academic Press, 2009).

vi. R.K. Sinha, K. Prasad, G. Sharma and R. Dalwani, 'Ecological Restoration of the River Ganga', *International Journal of Ecology and Environmental Sciences*, vol. 27 (2001): 127–35.

vii.   Raju Vyas, 'Mugger (*Crocodylus palustris*) Population In and Around Vadodara City, Gujarat State, India', *Russian Journal of Herpetology*, vol. 17, no. 1 (2010): 43–50.

viii. Raju Vyas, 'Current Status of Marsh Crocodiles (*Crocodylus palustris*) (Reptilia: Crocodylidae) in Vishwamitra River, Vadodara City, Gujarat. India', *Journal of Threatened Taxa*, vol. 4, no. 14 (2012): 3333–41.

ix.   Raju Vyas, 'Results of the 2015 Mugger Crocodile (*Crocodylus palustris*) Count at Vadodara, Gujarat, India', *IRFC Reptiles and Amphibians*, vol. 25, no. 1 (2018): 20–25.

x.   Ravindra Kumar Sinha, Sunil Kumar Verma and Lalj, Ganges River Population Status and Conservation of the Ganges River Dolphin (*Platanista gangetica gangetica*) in the Indian Subcontinent', in *Biology, Evolution and Conservation of River Dolphin within South America and Asia* (Nova Science Publishers, Inc., 2010).

xi.   Shin-ya Ohba, 'Ecology of Giant Water Bugs (Hemiptera: Heteroptera: Belostomatidae)', *Entomological Science*, vol. 22, no. 1 (2018): 6–20.

xii.   William Chambers and Robert Chambers, 'The Calcutta Adjutant', *Chamber's Journal of Popular Literature Science and Arts*, vol. 15, no. 368 (1861): 40–41.

xiii. N. Kelkar, B. D. Smith, M. Z. Alom, S. Dey, S. Paudel, and G.T. Braulik, *Platanista gangetica*. The IUCN Red List of Threatened Species (2022): e.T41756A50383346 (Online).

16. Cooperation or Conflict? Wars over Water

i.   Dhruv Sen Singh, *The Indian Rivers: Scientific and Socio-Economic Aspects* (Springer, 2018).

ii. Eric Keels, 'Praying for Rain? Water Scarcity and the Duration and Outcomes of Civil Wars', *Defence and Peace Economics*, vol. 30, no.1 (2017): 27–45.

iii. J. Liu, H. Yang, S.N. Gosling, M. Kummu, M. Flörke, S. Pfister, N. Hanasaki, Y. Wada, X Zhang, C. Zheng, J. Alcamo and T. Oki, 'Water Scarcity Assessments in the Past, Present and Future', *Earths Future*, vol. 5, no. 6 (2017): 545–59.

iv. Lyla Mehta, 'Contexts and Construction of Water Scarcity', *Economic & Political Weekly*, vol. 38, no. 48 (2003): 5066–72.

v. Malin Falkenmark, Jan Lundqvist and Carl Widstran, 'Macro-Scale Water Scarcity Requires Micro-Scale Approaches: Aspects of Vulnerability in Semi-arid Development', *Natural Resources Forum*, vol. 13, no. 4 (1989): 258–67.

vi. S. Janakarajan, M. Llorente and Marie-Hélène Zérah, *Urban Water Conflicts in Indian Cities: Man-made Scarcity as a Critical Factor* (UNESCO Working series SC-2006/WS/19, 2006).

17. Bengaluru: Landlocked City of Tanks and Lakes

i. Harini Nagendra, *Nature in the City: Bengaluru in the Past, Present and Future* (Oxford University Press, 2016).

ii. Hita Unnikrishnan and Harini Nagendra, 'Privatizing the Commons: Impacts on Ecosystem Services in Bangalore's Lakes', *Urban Ecosystem*, vol. 18, no. 2 (2015): 613–32.

iii. Hita Unnikrishnan and Harini Nagendra, 'Quenching a City's Thirst: The Shifting Waters of Bangalore', in *Encyclopaedia of the World's Biome* (Elsevier, 2020).

iv. Seema Mundoli, B. Manjunatha and Harini Nagendra, 'Lakes of Bengaluru: The Once Living, but Now

Endangered Peri-urban Commons', Azim Premji University, Working Paper Series, no. 10 (2020).

19. Dealing with Climate Change

i. Dick Eckstein, Vera Kunzel and Laura Schafer. *Global Climate Risk Index 2021: Who Suffers Most from Extreme Weather Events? Weather Related Loss Events in 2019 and 2000–2019* (GermanWatch, 2021).

ii. K. Vinke, H.J. Schellnhuber, D. Coumou, T. Geiger, N. Glanemann, V. Huber, J. Kropp, S. Kriewald, J. Lehmann, A. Levermann and A. Lobanova, *A Region at Risk: The Human Dimensions of Climate Change in Asia and the Pacific* (Asian Development Bank, 2017).

iii. M.K. Roxy, Subimal Ghosh, Amey Pathak, R. Athulya, Milind Mujumdar, Raghu Murtugudde, Pascal Terray and M. Rajeevan, 'A Threefold Rise in Widespread Extreme Rain Events over Central India', *Nature Communications*, vol. 8, no. 708 (2017): (Online).

iv. Navroz K. Dubash, *India in a Warming World* (Oxford University Press, 2020).

v. R. Krishnan, J. Sanjay, Chellappan Gnanaseelan, Milind Mujumdar, Ashwini Kulkarni and Supriyo Chakraborty, *Assessment of Climate Change Over the Indian Region* (Springer, 2020).

vi. Stefan Rahmstor, 'Rising Hazard of Storm Surge Flooding', *Proceedings of the National Academy of Sciences,* vol. 114, no. 45 (2017): 11806.

vii. Svante Arrhenius, 'On the Influence of Carbonic Acid in the Air upon the Temperature of the Ground', *Philosophical Magazine and Journal of Science,* vol. 5, no. 41 (1896): 237–76.

viii. UNESCO and UN-Water, *United Nations World Water Development Report 2020: Water and Climate Change* (UNESCO, 2020).

20. Guwahati: City of Beels and Landscaped Riverfronts

i. Anwesha Borthakur and Pradeep Singh, 'India's Lost Rivers and Rivulets', *Energy, Ecology and Environment*, vol. 1, no. 5 (2016): 310–14.

ii. B.K. Bhattacharjya, B.J. Saud, S. Borah, P.K. Saikia and B.K. Das, 'Status of Biodiversity and Limno-chemistry of Deepor Beel, a Ramsar site of International Importance: Conservation Needs and the Way Forward', *Aquatic Ecosystem Health and Management*, vol. 24, no. 4 (2021): 64–74.

iii. Chitrini Mozumder and Nitin K Tripathi, 'Geospatial Scenario-based Modelling of Urban and Agricultural Intrusions in Ramsar Wetland Deepor Beel in Northeast India Using a Multi-layer Perceptron Neural Network', *International Journal of Applied Earth Observation and Geoinformation*, vol. 32 (2014): 92–104.

iv. Edward Albert Gait, *A History of Assam* (Thacker, Spink & Company, 1906).

v. Samudra Gupta Kashyap and Rahul Karmakar, *Forever Guwahati* (Guwahati Metropolitan Development Authority, 2014).

vi. J.N. Sarma, 'An Overview of the Brahmaputra River System', in *The Brahmaputra Basin Water Resource* (Springer, 2004).

vii. John M'Cosh, *Topography of Assam* (Gyan Publishing House, 1837).

viii. Jyotisikha Dutta and Archana Sharma, 'Valuing Fishing Activity of the Deepor Beel', *Space and Culture, India*, vol. 7, no. 4 (2020): 122–32.

ix. Karobi B. Saikia, 'Threat to the Migratory Avian Fauna of Deepor beel: A Ramsar Site in Assam', *The Ecoscan*, vol. 7 (2013): 31–36.

x.   Pallabi Borah and Abani Kumar Bhagabati, 'Effect of River Environment on the Land Use of Guwahati City: Perspectives from Nature–Culture relationship', *The Clarion*, vol. 4, no. 1 (2015): 27–33.

xi. Rajib Gogoi, 'Conserving Deepar Beel Ramsar Site, Assam', *Current Science*, vol. 93, no. 4 (2007): 445–46.

xii. Renu Desai, Darshini Mahadevia and Aseem Mishra, 'City Profile: Guwahati. Centre for Urban Equity', Working Paper 24, (2014).

xiii. Surya Kumar Bhuyan, *Lachit Barphukan and His Times* (G.S. Press, 1944).

xiv. TERI, *Risk Assessment and Review of Prevailing Laws, Standards, Policies, and Programmes for Climate Proof Cities: Synthesis Report for Guwahati* (The Energy and Resource Institute, 2013).

21. Healing Waters and Holy Spirits

i.   Anna Perdibon, 'Between the Mouth of the Two Rivers: The Agency of Water, Springs, Rivers and Trees in Ancient Mesopotamian Cosmology and Religion', *Etnološka Tribina: Yearbook of the Croatian Ethnological Society*, vol. 51, no. 44 (2021): 34–53.

ii. Chetan Sahasrabudhe and Anurag Kashyap, 'Festivals, Rituals and Urban Landscape in Eighteenth-century Maharashtra', *South Asian Studies*, vol. 32, no. 2 (2016): 155–65.

iii. Githa Badikilaya, 'Chhath Puja: A Study in Religious and Cultural Tourism', *Atna Journal of Tourism Studies*, vol. 14, no. 2 (2019): 71–76.

iv. Harini Nagendra, *Restoration of the Kaikondrahalli Lake in Bangalore: Forging a New Urban Commons* (Kalpavriksh, 2016).

v.   M.U. Ushadevi, 'Famine in Folklore: The Case of the State of Mysore', *Proceedings of the Indian History Congress*, vol. 61, no. 1 (2001): 848–51.

vi.   Rana P.B. Singh, Pravin S. Rana and Sarvesh Kumar, 'Sacred Water Pools of Hindu Sacredscapes in North India', *Etnološka Tribina: Journal of Croatian Ethnological Society*, vol. 44, no. 51 (2021): 12–33.

vii. Rana P.B. Singh, 'Muslim Shrines and Multi-Religious Visitations in Hindus' City of Banaras, India: Co-Existential Scenario', in *Pilgrims and Pilgrimages as Peacemakers in Christianity, Judaism and Islam* (Routledge, 2014).

viii. Richard Laster, Rabbi David Aronovsky and Dan Livney, 'Water in the Jewish Legal Tradition', in *The Evolution of the Law and Politics of Water* (Springer, 2009).

ix.   Terje Oestigaard, *Water and World Religions. An Introduction* (SFU & SMR, 2005).

22. Pollution and Restoration: The Eternal Cycle

i.   Anil Agarwal and Sunita Narain, *Dying Wisdom: Rise, Fall and Potential of India's traditional Water Harvesting Systems* (Centre for Science and Environment, 1997).

ii.   Harini Nagendra, *Restoration of the Kaikondrahalli Lake in Bangalore: Forging a New Urban Commons* (Kalpavriksh, 2016).

iii. Okapi and Care Earth Trust, *Wetlands Restoration: A Comprehensive Handbook. Learnings from Chennai* (Okapi and Care Earth Trust, 2021).

iv.   Tony Joseph, *Early Indians: The Story of Our Ancestors and Where We Came From.* (Juggernaut, 2018).

23. Lakshadweep: Water Everywhere, Nor Any Drop to Drink

i.   DPP, *Union Territory of Lakshadweep: Basic Statistics 2014* (Directorate of Planning and Statistics [DPP], 2015).

ii.   K. Md Najeeb and N. Vinayachandra, 'Groundwater Scenario in Lakshadweep Islands', *Journal Geological Society of India,* vol. 78 (2011): 379.

iii.   R.H. Ellis, *A Short Account of the Laccadive Islands and Minicoy* (Asian Educational Services, 1924).

iv.   Sibin Antony, Vinu V. Dev, S. Kaliraj, M.S. Ambili and K. Anoop Krishnan, 'Seasonal Variability of Groundwater Quality in Coastal Aquifers of Kavaratti Island, Lakshadweep Archipelago, India', *Groundwater for Sustainable Development*, vol. 11 (2020): (Online).

v.   Stella James and Rohan Arthur. 'Breaching the Bounds: Urbanisation on a Low-lying Coral Atoll', in *Urban Sustainability Challenges in India* (Orient Blackswan, forthcoming).

vi.   Union Territory of Lakshadweep, *Lakshadweep Plan on Climate Change* (Department of Environment and Forestry, Union Territory of Lakshadweep, 2012).

vii.   Vijay Shankar Singh, *Evaluation of Groundwater Resources on the Coral Islands of Lakshadweep, India* (Springer, 2017).

ACKNOWLEDGEMENTS

It was a pleasure to work with Penguin Random House once again. We owe special thanks to Manasi Subramaniam, who stimulated the initial conversations that led to the development of this book. We also thank others at Penguin: Shubhi Surana, who worked closely with us, Manali Das for her careful edits, Swati Chandak Sharma for the illustrations, Gunjan Ahlawat for the book's superb design and Naina Tripathi for her assistance with publicity.

We are especially grateful to Azim Premji University for funding the research that went into the book and providing a stimulating professional environment in which to exchange ideas, and to write. The staff at the university library were especially helpful, assisting us in sourcing books and digging out obscure references—we owe them a special debt. Our heartfelt thanks to Abhiri Sanfui, Amrita Sen, Ayushi Chauhan, Dechamma C.S., Hita Unnikrishnan, Kulbhushan Suryawanshi, Manjunath B., Ravi Jambhekar, Sreerupa Sen and Sukanya Basu, our collaborators on research over the years, much of which has found its way into this book. We are grateful to Hita Unnikrishnan and Arvind Lakshmisha for their comments on early drafts of the book, to Rohan Arthur for his comments on the chapter on Lakshadweep, and Sharen Achangadan for collating the references. Finally, we thank the water warriors described in the pages of the book for sharing their inspiring journey with us.

Above all, we could not have written this book without help from our friends and families, and each of us has our own personal acknowledgements to add.

Harini

I am indebted to my family for sharing with me their love for nature: especially my mother Manjula Nagendra, my grandmother Thungabai and my mother-in-law Annapurna Sastri. To my father C.V. Nagendra, and my father-in-law S.J. Sastri (Appa and Baba)—constant champions of all my projects, you would have been so pleased to see yet another book taking shape. I miss you more than I can say.

Lakshmi's steadfast support was essential for me to write the book—she has my eternal gratitude.

Most importantly, this book could never have been written without the love and encouragement of Venkatachalam Suri, who has accompanied me on journeys of discovery for the past thirty years, and Dhwani Nagendra Suri, my biggest cheerleader. They inspire me and keep me going.

Seema

As a child, the highlight of my summer vacations to Kerala was swimming in the pond in my mother's ancestral house or in the Manali River on whose banks my father's ancestral residence stood. The many hours spent in the water trying to catch fish with towels and playing made-up games in the water with cousins while keeping a wary eye out for checkered keelbacks are treasured memories. I am very grateful to my parents, Savithry and Narayanan, who gave me these experiences by unfailingly making the long journey to Kerala every year, with my brother and me in tow.

Madhavi and Ram—thank you so much for your love and support especially during some rough patches, and the happiness the trips with you have brought. Thank you, Priyambad Pattanayak for agreeing to take my author photograph for this book, and your patience with an impatient subject. Prema Naraynen, friend of many decades, your advice and help with cat care has helped me get through some stressful times. Pratima Banka, thank you for the conversations, cat care and car rides. Lucky Begum, my heartfelt gratitude, for taking care of my home and my cats as if they are your own.

My aunts and uncles, Girija Neelakandhan, M.N. Neelakandhan, Sreedevi C.P., M.N. Vasudevan and M.N. Devaky, I cannot thank you enough for your support in a year of crises.

Venku and Ravi, the two men in my life without whom whatever I do is impossible. Venku—thank you for being a calming influence on me. I could not have asked for a more supportive partner and companion. Ravi—thank you for the happiness of the summer afternoons of our childhood, and for every day of your affection since that keeps me afloat.

Much of this book was written with a bundle of foster kittens on my lap. These orphaned (and often ill) kittens have contributed invaluably to this book with their insistence on sleeping on my lap, playing on the desk or pouncing on the keyboard, while I worked. Noori, a discard of the cruel kitten breeding mill, who came to me starved and sick, thank you for trusting me to care for you. My resident felines Mykah and Fuzzy, whose paws I have held for comfort, you are the purest joy of my life—thank you.